Cultures of Milk

Cultures of Milk

The Biology and Meaning of Dairy Products
in the United States and India

ANDREA S. WILEY

 Harvard University Press

Cambridge, Massachusetts
London, England
2014

Library of Congress Cataloging-in-Publication Data

Wiley, Andrea S., 1962–   author.
   Cultures of milk : the biology and meaning of dairy products in the United States
and India / Andrea S. Wiley.
      pages   cm.
   Includes bibliographical references and index.
   ISBN 978-0-674-72905-6 (alk. paper)
      1. Milk—Social aspects—United States.   2. Milk—Social aspects—India.
3. Milk—United States—History.   4. Milk—India—United States.   5. Dairy
products—United States—History.   6. Dairy products—India—History.
7. Food preferences—United States.   8. Food preferences—India.   I. Title.
[DNLM: 1. Milk—India. 2. Milk—United States. 3. Cross-Cultural Comparison—
India. 4. Cross-Cultural Comparison—United States. 5. Socioeconomic Factors—India.
6. Socioeconomic Factors—United States.]
   GT2920.M55W54   2014
   394.1'2—dc23   2013043044

For Himangi and Lalita, for your friendship

Contents

Preface and Acknowledgments

This work grew out of two strands of my intellectual interests. I have long been interested in cow milk as a human food, which has always struck me as an incongruous pairing. Mammalian milk is produced to serve particular growth and development needs among young bovines, and the question of how cow milk's biology impacts human biology (for better or for worse) has been a central preoccupation of my research for the past 10 years. Furthermore, one of the best-documented examples of human biological variation observable across populations is in the ability to digest milk in adulthood. Clearly, experiences with domesticated animals used for dairying has had a profound impact on the biology of human societies that made use of the milk of other mammals. As such, milk provides an ideal material for the study of biocultural interactions, which is the common underlying concern of all of my research. I have been intrigued by the complex interrelationships between human biology and human culture ever since I discovered anthropology as an undergraduate student, and have always found that food represents a compelling means to explore such connections. In the case of milk, not only does milk from any other nonhuman mammal have the potential to alter human biological function, but beliefs about its biology have contributed to choices people make about its consumption, or whether they compel themselves or their children to drink it.

There is less biology and more culture in this book than in my previous writings on milk. It represents my other scholarly interest, which has been in South Asia. My initial forays into work on milk were in the United States, whose "culture of milk" I felt quite familiar with. I had grown up with assumptions about milk's benefits and took for granted its essential role in the diet; I dutifully and unquestioningly drank the carton of milk placed on my school lunch tray. Conducting a study on milk consumption and child growth in India was an opportunity to explore another major dairy culture of the world, and in doing the research for this book it became clear that India's dairy culture (especially as it related to consumption rather than production) had been little explored by social scientists. India and the United States are the two largest producers and consumers of milk (the European Union notwithstanding, which outranks both of them). The two cultures have interacted indirectly around milk for millennia, with related cow domesticates and human mutations for lactase persistence. In the late twentieth century these interactions became direct when milk surpluses from the United States, along with those from Europe, became the foundation for the "White Revolution" in India under the guise of Operation Flood. Showing the entanglements of the two countries around milk is one goal of the book; situating milk in their separate histories and sociocultural contexts is another. The book revolves around those factors that shape consumption; I have conveniently sidelined the large literature on milk production in India except insofar as it influences consumption.

The sacred cow trope is one way to illuminate these similarities and differences. Most readers will be familiar with the fact that the cow is sacred to India's majority Hindu population. This status has come to mean injunctions against its slaughter or consumption, and for some it kindles puzzlement over the apparent "irrational" attachment to the cow. But as I hope to demonstrate, there are plenty of ways in which the cow attains something akin to sacrosanct status in the United States, with the dairy and beef industries vigorously defending the products of their cows against any challenge that would decenter these foods from American meals. Milk is privileged in different ways, from the way cows and/or the dairy industry figure in political and economic debates to how consumers or public health nutritionists articulate its impacts on health, especially that of children. Milk figures into multiple discourses in public and private, religious and secular life, in part through the imagined effects on the body, which I maintain are, in some ways, related to milk's evolutionary function as the sole food for rapidly growing infants.

I do want to acknowledge the limitations of this work. While there are several scholarly books on the history of milk consumption in the United States, most are concerned with the nineteenth and twentieth centuries, and there is no such published literature on South Asia. My primary question for the historical chapters of this work (Chapters 2 and 3) was how milk and milk products were consumed in the past. I have relied heavily on secondary sources for those chapters, and the material is more descriptive than analytical. Secondly, while I am an anthropologist and have traveled and conducted research in India over the past twenty years, I am not an ethnographer. My work has focused on various aspects of child growth and understanding the environmental (including dietary) factors that influence it. Thus experts on South Asia may wish for a more thorough analysis of themes I cover only in modest ways here, such as gender, religion, or village life. My hope is that this work might stimulate the interest of others in South Asia to delve deeper into its milk culture.

This project would not have gotten off the ground without the financial support of the Wenner-Gren Foundation. A post-PhD research grant supported my research trips to India in 2010, which were essential to the material presented here on maternal perceptions of milk in relation to their children's growth and development. The Government of India graciously provided me with a research visa to conduct the research. Indiana University, Bloomington awarded me a sabbatical in fall 2011 to support the writing of the book, which I had the good fortune to work on while ensconced at the Bodleian Library in Oxford. My thanks to Oxford University for allowing me access to their collections, to Harry West for hosting me at the School for Oriental and African Studies Food Studies Centre in London, and to the Leverhulme Centre for Integrative Research in Agriculture and Health, which provided me much-valued office space.

I am deeply indebted to my friends and colleagues at King Edward Memorial Hospital Research Centre Diabetes Unit. Himangi Lubree, Lalita Ramdas, Dattatray Bhat, Suyog Joshi, Charu Joglekar, and most especially Dr. C. S. Yajnik welcomed me to their research unit and made this work possible. Many others there also facilitated my research, and I apologize for not thanking everyone by name. I am grateful to all for their collegiality, support, and friendship.

I would like to thank my editor at Harvard University Press, Michael Fisher, for encouraging me to write the book and of course enabling its publication. I am most appreciative of Janet Chrzan and John Allen, both of whom read the manuscript draft and provided helpful comments that have substantially strengthened the book. I take full responsibility for any and all shortcomings that remain.

In Bloomington, Diane Richardson has made juggling teaching, research, and administrative work so much easier, and I appreciate that more than I can ever express. Lindsey Gooden Mattern helped enormously with the initial stages of library research on milk in India, while Virginia Vitzthum provided steadfast moral support and friendship throughout. Most especially I am grateful to my family—Rick, Aidan, and Emil—for supporting my work and forgiving my lapses of attention while I was immersed in the writing.

Cultures of Milk

1

Introduction
Cultures of Milk

Milk. Cheese. Yogurt. These products have been a mainstays of the diets of northern Europeans since cattle were domesticated in the region about 8,000 years ago. The proportions of consumption of each have changed over time, most notably with the rise of fresh milk consumption in the nineteenth and early twentieth centuries, but it is fair to say that when most people think about dairy products they are envisioning them as part of European cuisines, produced by placid large-bodied black-and-white Holstein or fawn-colored Jersey cows grazing in verdant pastures, and consumed by tall, robust, light-skinned people. As a representation of a global pattern of milk production and consumption, this image was never quite accurate. While northern Europe certainly has a well-known set of dairy traditions underpinned by such cows and other similar breeds, this region is not the only one to have a long-established set of practices around milk production and consumption. Nor have northern Europeans always been tall and robust.

Nomadic populations of central Asia, as well as East and West Africa, rely on milk and milk products as dietary staples, and herd a variety of animals (cows, horses, camels, goats, and sheep). But it was in South Asia, among large-scale settled populations that grew crops and kept domestic animals, where the other major dairy culture of the world developed. Here milk comes mostly from another large bovine species, the water buffalo *(Bubalus bubalis),* as well as local cows of the species *Bos indicus,* as opposed to *Bos taurus* in Europe. Goat milk is also drunk, although not as extensively and mostly in rural areas. In fact India currently produces and consumes more milk than any other country, although given its large population size, ecological diversity, and high poverty rates, both per capita production and consumption are relatively low. Furthermore, the way in which milk and milk products are consumed is quite different from Europe and among its derived populations in North America, Australia, and New Zealand.

Aged cheese, for example, a quintessential dairy product in areas populated by Europeans, has never been part of the portfolio of Indian dairy products. Fermented milk and production of curds (yogurt) have a deep history in India, but cheeses created from active acid-based separation of curds from whey are largely products of the Portuguese colonial enterprise. Fresh cheeses such as *chhana* and *paneer* now have a variety of culinary uses, but commercial production is limited and these are not savored on their own, but rather incorporated into sweetmeats and curries. Given the centrality of elaborate cheese crafting to European dairying traditions, its absence in India may have contributed to a lack of international attention to indigenous Indian dairy customs. Even when the British ruled much of the subcontinent, there was little appreciation for Indian dairy products. It is also the case that India has not exported its dairy products—aged cheeses are well suited to the global marketplace in ways that fresh dairy products are not—and demand for milk has exceeded domestic supply in India. European Union countries are the major players in the global dairy trade, with Australia, New Zealand, and the United States having more regionally bounded trade networks for dairy.

Understanding Similarity and Difference in Dairy Cultures

In this book I bring Indian dairying traditions to the fore, and compare and contrast them with those in the United States, which are derived almost exclusively from Europe, especially northern Europe. I have conducted research on milk in both countries, and I find that insights into both dairy cultures can be enhanced by comparisons between them. I am less interested in the specifics of dairy production and more concerned with the ways in which various historical, cultural, and political economic forces have shaped how people think about milk and milk products and how these influence consumption in two large countries with robust dairy cultures.

India and the United States have long placed high value on these commodities but each has unique ecological, geographical, historical, and cultural circumstances. While the rise of fresh milk consumption in the United States and northern Europe is reasonably well documented, the fluctuating cultural status of milk and other dairy products has not been the subject of much scholarly work. Moreover, this and related consumption practices in India, with its large-scale dairy industries, have not been systematically studied. How milk and milk product consumption and Hindu ideals about the sanctity of the cow articulate with the political economy of milk production and consumption and how India negotiates the status of milk in ways that are similar to and/or different from the United States, where cows are—for all practical purposes—the only source of milk, is a major theme.

Throughout the book I rely on material from ethnographic work, literature, popular media, advertisements, official policies, published scientific work, and my studies of milk consumption and child growth, among other types of data, in order to address these questions. I consider the range of variability within and between each country with respect to uses of milk, the physiology of milk digestion (i.e., adult lactase production), and public health policies related to diet. I also examine the varying importance of nonfluid forms of milk (emphasizing butter, *ghee,* cheese, and yogurt [*dahi*]); complexities associated with different types of milk (cow and water buffalo in India; cow and plant-based milks in the United States); and the relationship between children and milk, including marketing and nutrition education practices as well as evidence concerning milk and child growth.

The overarching approach I take to this investigation is a biocultural one (Wiley, 1992, 1993; Wiley and Allen, 2013), which analyzes the interplay between evolutionary processes, biological outcomes, and social and cultural institutions and ideals. In this case, I assess variation and similarity in milk consumption and the meanings attached to milk within the two countries in relation to their evolutionary and social histories with milk. This deep historical and biological foundation is then considered in light of more recent and contemporary ecological, economic, and political processes that established bovine domestication, dairying, and dairy consumption. I focus on the ways in which these have changed over the twentieth and twenty-first centuries and affect contemporary beliefs and practices related to milk consumption. Developments at the local, national, and global levels influence milk consumption in India and the United States and thereby contribute to similarity and variability in the dairy cultures of the two countries.

I start from the observation that, as a food, milk possesses some unique qualities, and that these have been manipulated and understood within different social and cultural contexts. Thus milk's material nature is front and center, along with its known and imagined effects on the human body. As Peter Atkins concluded in his work on milk's shifting qualities as it was subject to biochemical and legal scrutiny in the nineteenth and twentieth centuries in the United Kingdom, "The everyday material of our lives, including food and drink, because it is unconsidered, because it is unchallenged in its significance, is a powerful means of guiding our expectations—in the case of food, our habituated, embodied norms of nutritional sufficiency and bodily reproduction" (Atkins, 2010, 279). The ideas that people hold about milk's qualities shape its consumption; these have their own history, which I maintain are derived in significant ways from milk's biology and hence its perceived effects on human biology, especially that of children.

Debates about whether it is more useful to interpret consumption of a particular food as adaptive under a set of ecological and subsistence conditions or if food production and consumption practices are themselves strongly influenced by ideological systems have a long history in anthropology. These competing modes of explanation are best represented by debates over pork taboos among Jews and Muslims, cannibalism among the Aztecs, and prohibitions on cow slaughter and beef consumption among Hindus (cf. Douglas, 2002 [1966]; Harner, 1977; Harris, 1985; Ortiz de Montellano, 1978; Sahlins, 1976). Materialist perspectives—and evolutionary perspectives from the biological sciences more generally—consider adaptive significance (i.e., whether the behavior enhances survival and/or reproduction) as primary and maintain that the benefits of a given behavior should outweigh any costs, measured in economic or Darwinian fitness terms (Brown et al., 2011; Harris, 1979). Normative views follow from that.

On the other hand, both Mary Douglas and Marshall Sahlins argue that ideals are primary, and that these in turn drive behavior, such as what to eat or what to produce. Sahlins articulated this view in the following: "the world environment [of production of agricultural commodities such as meat] is organized by specific valuations of edibility and inedibility, themselves qualitative and in no way justifiable by biological, ecological, or economic advantage" (Sahlins, 1976, 171). The debate between these two rather entrenched perspectives figures prominently in the question of why Hindus consider the cow to be sacred. Marvin Harris argued that it is the economic contributions of the cow that protect it from slaughter (Harris, 1966). This perspective generated a sizeable counterliterature that attempted to show that other factors—Hindu scripture, cosmology, political agendas—underpinned the cow's sacred status, or provided data that undermined the claim of economic value (see Chapter 3).

In my view, there is nothing inherent to these perspectives that renders them mutually exclusive. A biocultural approach allows for the ways in which a particular behavior (e.g., consumption of milk) might be considered adaptive and how the normative views that people hold influence that practice. Ideals, worldviews, and principles have histories and don't act as autonomous or ahistorical motivators of behavior; instead these—and their ability to influence behavior—are also in flux in relation to eco-logical, economic, political, demographic, and other social processes (Wiley, 1992). With regard to milk, how this food figures in people's imaginations is important. What people think milk ingestion will do for them, or what drinking—or at least purchasing—milk means to them motivates those practices, at least in a proximate sense. In turn, the meanings of those biological effects, as well as manifest biological outcomes, are also critical. In both the United States and India milk fits into conceptual-izations of what the body *should* look like and in fact may influence what the body *does* look like.

To use a related example, Sidney Mintz demonstrated that sugar fit into new conceptualizations of time and meal patterns framed by industrial work as it infiltrated the British diet during the nineteenth century (Mintz, 1985). Sugar became the energy source for the industrial proletariat, with deleterious consequences for the diet. I argue that milk may have served to rectify some of the nutritional shortcomings of this dietary shift by becoming the food associated with the modern body—due to its per-ceived abilities to make it bigger, stronger, and more powerful—and these qualities extended to the nation-state (Wiley, 2011a). Of course, milk and sugar became mingled in tea (or *chai* in India): both tea and sugar were colonial commodities, grown in different hemispheres on a large scale. They flavored and sweetened a familiar domestic commodity, although it is important to keep in mind that fresh milk consumption was not neces-sarily routine in the United Kingdom, the United States, or India. In my view, milk consumption was bolstered by its use in this new bittersweet beverage, as well as its colonial cousins, coffee (which replaced tea in the United States as the vehicle for sugar and milk) and chocolate.

Why Milk?

Why devote an analysis to milk and milk products? In what way is this topic and comparison more compelling than, say, a comparison of other foods the regions have in common, such as wheat (leavened bread versus flatbreads), or legumes (peas versus lentils)? Insofar as the close analysis of any given food is a lens into different cultures and the historical pro-cesses that have shaped culinary traditions, there is no real difference. On the other hand, in both places milk is marked as a "special" food, one with attributes that have no equivalent in other foods. Because of this, milk is particularly elaborated in popular discourse and embedded in political, economic, cultural, and health discussions and institutions. Furthermore, milk is often considered to have specific consequences for the biology of its consumers, so it tells us much about how people think about their bodies, and how their bodies in turn signify something important about their social lives. As Sidney Mintz wrote in *Sweetness and Power: The Place of Sugar in Modern History,* a book that has become a touchstone for food scholarship, "In understanding the relationship between commodity and person, we unearth anew the history of ourselves" (1985, 214).

There are several ways in which milk is considered special, and these stem from recognition of its unusual source. Milk is produced by maternal mammals as the initial food for their infants. Indeed, milk production by females is a defining trait of this class of animals. What this means too is that milk is the only food that is produced in order to be consumed (honey is another, though it is produced from pollen rather than from bees themselves). The rest of the food that mammals, including humans, eat comes from some form of predation—that is, we have to "kill" plants and animals in order to eat them. Of course, milk is produced by mammals only for consumption by infants, for whom it is their sole food for some length of time. As such, milk is tailored to meet the growth, developmental, and immunological needs of infants of a given species. Given its role in this regard, milk is often described as a "complete" food (blurring the "food/drink" dichotomy), but what constitutes a "complete" food for infants is likely to be quite different from what older children or adults should consume. It is also the case that mammalian milks are highly variable in their composition, and milk from one would be quite unsuitable for another, as each species has different growth rates, sizes at birth and adulthood, and ecological niches (Oftedal and Iverson, 1995). For example, cow milk is too low in iron to be suitable as a breastmilk substitute for young human infants, and it is much richer in protein and calcium to support the rapid growth of a large bovine skeleton.

Humans are unique insofar as many people drink the milk of other mammalian species (mostly bovine), and they consume it well beyond the age of weaning. As I already noted, this has not been the historical norm for humans. The production and consumption of milk and milk products has been restricted to nomadic pastoralist groups and settled societies of Europe and South Asia. Prior to European colonization, there were virtually no domesticated mammals and certainly no dairying traditions among Native American, Australian, or Oceanic populations. In East and Southeast Asia, the use of domesticated animals for dairying was more or less nonexistent. Dairying played a large role in Tibetan and Mongolian diets, and during Mongolian control of China there was some dairy presence in the diet, but subsequently usage of milk products died out (Anderson, 1987; Huang, 2002). In Sub-Saharan Africa, very clear cultural and dietary distinctions existed between pastoralists, who relied heavily on their dairy animals, and agricultural populations, who made little or no use of dairy products (Simoons, 1981). European dairy traditions took hold across North America, Australia, and New Zealand, and these regions stand out in the current global distribution of milk availability, as illustrated in Figure 1.1. Despite being the largest milk producer in the world, India is not prominent in Figure 1.1, as overall per capita levels are low.

Historically there was geographic overlap between populations who practiced dairy production and consumption, those with the ability to digest the milk sugar lactose throughout life, and those who highly valued milk and milk products. In his widely read set of essays *Good to Eat: Riddles of Food and Culture* (1985), the late anthropologist Marvin Harris was so impressed by this differentiation that he argued for a classification of human societies into the "lactophiles" and "lactophobes." For Harris, the predilection or distaste for milk as a food was a profound marker of cultural distinction, with the lactophiles forming a distinct minority. Not surprisingly, the lactophiles placed a high cultural value on milk and had high frequencies of the genetic mutation that allowed for digestion of the milk sugar lactose throughout life. Lactophilic societies

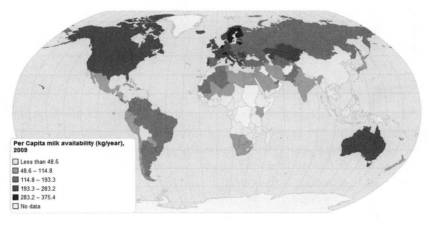

Figure 1.1 Current per capita global availability of milk for consumption.
Data from FAOSTAT, Statistics Division, 2013.

sang the praises of milk as "nature's perfect food," while lactophobes scorned it. As Harris wrote,

> The Chinese and other eastern and Southeast Asian peoples do not merely have an aversion to the use of milk, they loathe it intensely, reacting to the prospect of gulping down a nice, cold glass of the stuff much as Westerners might react to the prospect of a nice, cold glass of cow saliva. Like most of my generation I grew up believing that milk is an elixir, a beautiful white liquid manna endowed with the capacity to put hair on manly chests and peaches and cream on women's faces. What a shock to find others regarding it as an ugly-looking, foul-smelling glandular secretion that no self-respecting adult would want to swallow. (1985, 130–131)

In the main, India and the contemporary United States fit squarely into Harris's lactophilic category of human societies. Milk production and consumption exist in both countries within cultural frameworks that imagine milk in various ways and imbue this food with significance that goes well beyond its nutritive value. In both South Asia and the contemporary United States, milk is thought to have properties unmatched by other foods, with the ability to transform bodies in favorable, even miraculous, ways. Milk products such as various cheeses, yogurt, or butter appear to be conceived somewhat differently. At the same time, there are anxieties about milk, because it is vulnerable to rapid spoilage or contamination by life-threatening pathogens. It is easily adulterated or diluted by unscrupulous suppliers, placing consumers at risk from a highly valued food whose consumption is strongly encouraged. Thus there are numerous ways in which milk is both biologically and culturally unique as a food and considered "special." Furthermore, its biological properties are related to its cultural valuation, and how these are constructed and shift over time in the world's two largest milk producing and consuming countries is what I hope to illuminate.

Lactose and Lactase Persistence

Understanding how and why populations vary in rates of milk digestion has been a major preoccupation of anthropological geneticists, as it is one of the clearest examples of population-patterned genetic variation that can be clearly tied to differences in subsistence patterns. A brief discussion of this genetic trait is needed in order to understand how milk came to be widely consumed and appreciated in these two regions and the sources of this genetic heterogeneity in milk digestion in India and the United States. This trait further reveals milk's uniqueness as a food and the ways in which evolutionary processes related to dairying cultures played out in South Asian– and European-derived populations.

Population differences in use and perceptions of milk are related to milk's unique sugar, lactose.[1] Lactose is a double sugar (a disaccharide), made up of glucose and galactose, and cannot be absorbed in the small intestine directly. Instead it must be cleaved into these single sugars, which can then be absorbed, enter the body's circulatory system, and used for energy or converted into fat for storage. This initial process requires a specialized enzyme called lactase, which is found along the cells that line the upper small intestine. In general, infant mammals produce lactase in order to break down the lactose they ingest in their mother's milk. However, lactase production diminishes over time and eventually stops altogether, usually around the time of weaning. Importantly, mammals living in the wild never consume milk again after they are weaned; since the sole function of lactase appears to be its ability to cleave lactose, it would be wasteful of scarce energy and nutrients to continue to produce a useless enzyme.

Among humans, the gene for the lactase enzyme (abbreviated as LCT) is on chromosome 2, but it is a regulatory region upstream from the lactase gene itself that is the site of variation in adult lactase activity (Enattah et al., 2002; Ingram et al., 2007, 2009; Tishkoff et al., 2007). Some Sub-Saharan African, European, Middle Eastern, and South Asian populations have shown high frequencies of mutations that function to keep lactase activity throughout life, while all other humans have the ancient mammalian DNA sequence, which results in lactase being turned off around the time of weaning (see Figure 1.2, which shows the geographic areas of the Old World with high frequencies of lactase persistence). Note that the variation in the frequency of lactase persistence alleles generally corresponds to the variation in milk availability for consumption shown in Figure 1.1. The mutations that result in continued lactase production appear to be dominant to those that result in lactase being shut off around the time of weaning, and thus only one copy of the mutation is needed to continue to produce lactase into adulthood.[2] The decline of lactase production is independent of the presence of lactose in the small intestine (Sahi, 1994a, 1994b); continued milk drinking does not cause lactase production to remain, and abstaining from milk has no influence on lactase production. Individuals who continue to produce lactase in adulthood are said to be lactase persistent, while those whose lactase activity diminishes in childhood are lactase impersistent.

All groups that have high frequencies of the mutations for lifelong lactase activity share a history of dairy animal domestication and reliance

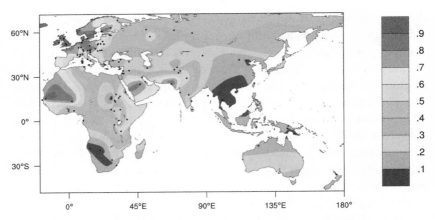

Figure 1.2 Distribution of the lactase persistence allele in Old World populations. © 2010 Itan et al., A worldwide correlation of lactase persistence phenotype and genotypes. *BMC Evolutionary Biology,* 10(1), 36; licensee BioMed Central Ltd. This is an open access article distributed under the terms of the Creative Commons Attribution License (http://creativecommons.org /licenses/by/2.0), which permits unrestricted use, distribution, and reproduction in any medium, provided the original work is properly cited.

on milk from these animals. Dairy animals were domesticated in the Old World around 8,000 years ago, and the mutation for ongoing lactase production appears to have spread after that time (Burger et al., 2007). It is easy to hypothesize that mutations that allowed people to digest milk throughout their lives would have been advantageous in dairying populations, as they would have been able to utilize a new nutrient-rich food. Individuals with the mutation would have been better nourished and presumably healthier, and ultimately had more offspring who also had the mutation (i.e., they would have had higher evolutionary fitness: Ingram et al., 2009; Simoons, 1978, 2001; Tishkoff et al., 2007). Within the context of this novel dairying environment natural selection would have caused this mutation to spread rapidly throughout these populations. Among human groups in the New World, East Asia, Oceania, or other areas of Africa who did not domesticate milk-producing animals or drink their milk, ongoing lactase production would have offered no advantage, since there was no milk in the diet after weaning. As a result, most humans have maintained the ancestral mammalian pattern in which lactase activity is turned off during the juvenile period.

Milk is certainly a nutrient-rich food, but it is also possible to gain nutritional benefits from it by using simple processing techniques that reduce the lactose in milk, thereby circumventing the need for continual lactase production. The most common means by which milk is processed around the world is to heat milk and then let it sit at warm temperatures, which encourages fermentation ("culturing") by lactophilic bacteria (such as *Lactobacillus*). This process converts lactose into lactic acid and yields yogurt (Mendelson, 2008). Depending on how long this process goes on and the quantity of bacteria present, most of the lactose will be removed, leaving the tangy taste of yogurt.

Alternatively, adding a strong acid such as lemon juice or vinegar, or the enzyme rennet, which is produced by the cells lining mammalian stomachs, to heated milk will cause it to curdle, as the milk solids ("curds") separate from the liquid whey portion. The curds can then be pressed, flavored, aged, or processed in other ways to make cheese. Lactose is dissolved in the liquid whey that is drained off; hence cheese has little lactose in it. In general, the softer the cheese, the more whey is retained in it, and the higher the lactose content. A very dry cheese such as Parmesan contains virtually no lactose; mozzarella, ricotta, *paneer*, or other fresh cheeses contain much more. In European traditions, butter is created from skimming off the cream in milk and contains relatively little whey, and hence little lactose. In South Asia, butter is made from yogurt. The cream layer is skimmed off and then cooked down to remove any remaining solids and water. This results in *ghee*, the clarified butter typically used in India, which has no lactose remaining.

Given that these are very basic processing techniques, and that they result in dairy products with little lactose but many nutrients left in them, why did the mutation for ongoing lactase production spread? It is worth noting that culinary behaviors can spread very rapidly through observation, teaching, and diffusion, while genetic change is relatively slow, with advantageous genes spreading through the population over many, many generations. One would expect that there must have been advantages to being able to consume milk in fluid form, or as whey, rather than exclusively as more solid dairy products with their lower levels of lactose. Indeed, William Durham (1991) observed that contemporary populations with high frequencies of adult lactase activity not only keep dairy animals, but they also drink *fresh* milk, which has the highest concentration of lactose. It remains unclear what the nutritional advantage would have been for consuming fluid, unfermented dairy products. One possibility is that fluid milk or whey serve as sources of water, which would have been important to pastoralist groups living in arid areas and relying heavily on their animals for food. Furthermore, animal products as a group are not a source of starch (sugar); milk is the only source for sugar (in the form of lactose) in an animal foods–based diet, and is the main fuel for the brain.

But this need for milk as a fluid or starch seems a less compelling explanation for populations in northern Europe and South Asia, where grains were also being cultivated and alternative sources of fluid (ample water or fermented beverages) were accessible. One proposal is that because northern Europe is characterized by low levels of UV-B light, which in turn is necessary for vitamin D synthesis, lactose, which is found only in fresh milk or whey, can perform one of the functions of vitamin D (Flatz, 1987; Flatz and Rotthauwe, 1973), which is to facilitate calcium absorption in the small intestine. Although this is an intriguing hypothesis, some researchers have found no evidence for an effect of latitude (as a marker for UV-B exposure and level of vitamin D synthesis) on the spread of mutations for adult lactose digestion in Europe (Itan et al., 2009), although others show that this possibility cannot be excluded (Gerbault et al., 2009). Whatever its significance to northern European populations, the vitamin D/UV-B light hypothesis has little relevance for understanding the spread of lactase persistence in South Asia.

Rates of lactase persistence are variable among contemporary populations of both the United States and India. In the former, prior to European colonization, Native Americans had no exposure to nonhuman milks, as there were no domesticated mammals used for this purpose. Virtually all Native Americans are lactase impersistent, while most European colonists who hailed from northern European countries are lactase persistent. Africans brought to North America as slaves are similarly lactase impersistent, as they mostly are derived from West African agricultural groups. Due to ongoing migration from southern Europe, Asia, and Latin America, as well as intermarriage among groups over the history of the United States, the overall rate of lactase persistence is estimated to be somewhere around 75 percent (National Dairy Council, 2012). This represents an average of different subpopulation rates: European Americans have rates around 90 percent (with higher rates among those with ancestry in northern Europe than those from the Mediterranean region), while African Americans, Native Americans, and Asian Americans have rates of around 10 to 25 percent. These estimates are based on small samples and may not represent the true frequency in these populations (NIH Consensus Development Conference, 2010).

India likewise varies in terms of lactase persistence rates, although there has been considerably less investigation of this phenomenon in that country compared to the United States. As Figure 1.3 indicates, the Indian pattern shows a gradual shift from high rates in the northwest to lower rates in the East and South. The most common mutation associated with lactase persistence is the same one found in high frequencies of European populations (Gallego Romero et al., 2012), and it appears that the allele for lactase persistence spread into South Asia from Europe as pastoralist populations moved east and south into the Indian subcontinent. A recent countrywide survey of the DNA segment where common variants in lactase activity are known to exist confirmed earlier reports of observed variation in lactose digestion (Gallego Romero et al., 2012). Most importantly, geographic location is the primary predictor of lactase persistence; linguistic affiliation also contributes to variation, with Indo-European language speakers having higher rates than Dravidian speakers. The former cluster in the North and West, with the latter distributed throughout South India. However, even after considering variation in language affiliation, the geographical pattern remains. The allele is particularly high among

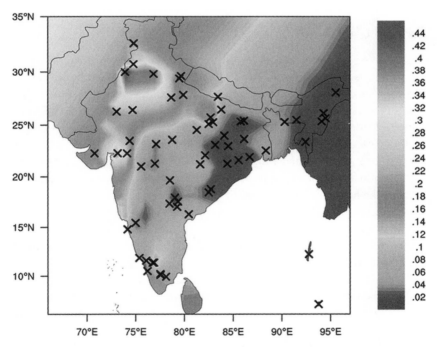

Figure 1.3   Distribution of the lactase persistence allele in South Asia. © 2011 Gallego Romero et al., Herders of Indian and European cattle share their predominant allele for lactase persistence. *Molecular Biology and Evolution,* 29(1), 249–260. Published by Oxford University Press on behalf of the Society for Molecular Biology and Evolution. All rights reserved.

pastoralist groups in India, highlighting its coevolution with dairying practices.

Nonetheless, alleles conferring lactase persistence exist in surprisingly low frequencies in India, with an average of about 13 percent and a range from 0 to 23 percent. The expression of lactase production in adulthood likewise ranges from 0 to 40 percent, with a countrywide average of about 20 percent (Gallego Romero et al., 2012), and it correlates with milk consumption (Baadkar et al., 2012). So, compared to population of northern Europe, where these percentages are over 90 percent, rates of lactase persistence are rather low, despite South Asia being a "dairy culture." This suggests that there was something about the dairying tradition and culinary importance of dairy products, historical processes, or the environment that shaped the distribution of lactase persistence differently in South Asia compared to Europe, even though both share Northwest–Southeast gradients in lactose digestion. The question of the role of milk in Indian prehistory and early history will be taken up in Chapter 3.

The Social Life of Dairy in India and the United States

Anthropological scholarship on milk has emphasized biological variation in milk digestion, and there is the related academic debate about reasons behind the cow's sacred status among Hindus. But milk figures in other discourses in each country, which provide insight into existing practices, normative ideals, and perceived problems related to milk. Scholarship and popular discourse about dairy reveals and reflects the quite distinct dairy cultures of India and the United States, and provides an indication of how dairy production or consumption is considered a topic worthy of engagement in the halls of research institutions, the glare of the public debate, or the intimacy of the household.

The two countries share some key features: both are large in land mass and population, and major milk producers. As noted, India now produces more milk than any country, and the United States is ranked third (the 27 countries of the European Union collectively outrank both). Not surprisingly, milk production is a cornerstone of their agricultural economies, comprising 11 percent of the U.S. economy and 17 percent of the Indian economy (Delgado and Narrod, 2002; U.S. Department of Commerce, 2002). Per capita intake is low in India relative to the United States (39 versus 95 liters per year, based on national availability data) due to constraints on access and a population of over one billion citizens, but overall consumption now exceeds every other country (U.S. Department of Agriculture, Economic Research Service, 2013). Milk is recommended in dietary guidelines produced by the U.S. Department of Health and Human Services and the U.S. Department of Agriculture, and by the National Institute of Nutrition in India.

Milk is venerated in both countries, in part through its status as a "pure" food, although the meanings of purity only partially overlap in both cultures. In India, milk's purity derives from being a product of the cow. Among India's majority Hindu population, milk's association with the cow transports it into the realm of the sacred. Cow milk is considered pure; foods cooked in it or individuals who consume it take on the sanctified qualities of the cow. Krishna (Figure 1.4), the popular childlike incarnation of the Hindu deity Vishnu, is well-known and beloved for his affection for cows, cow milk, and butter, all of which contributes to milk's position as a divine substance. The connection between milk and the "holy cow" is more complex, as most milk in fact comes from water buffalo (Figure 1.5), whose status is anything but sacred. Considered slow and dull-witted, these beasts are also iconographically associated with death. In practice there is slippage in the use of different kinds of milk for ritual purposes, with buffalo or goat milk often standing in for cow milk. However, buffalo milk is much richer in the prized butterfat, and it has greater economic value, leading to varied and sometimes competing discourses about the value of these and other types of milk in India.

Northern European colonists established the dairy culture of the United States quickly after their arrival. Milk production by cows and cheese and butter making featured centrally in the culinary practices of colonies in New England and down the East Coast. Cows became, and remain, for all practical purposes, the only source of milk. While cows have no official religious status, dairies do use biblical references to "milk and honey," or describe their cows as "divine bovines." Indeed there is an echo of Hindu views of the cow in Pennsylvania State University's dairy scientist Stuart Patton's comment in his contemporary paean on milk: "Perhaps we need to ponder what higher, better purpose the cow could have in this life. . . . The cow is a profound gift to humanity" (Patton, 2004, 168). Cows serve as icons of bucolic innocence, and they are often represented in milk promotions of various kinds, as well as figuring in wide-ranging styles of popular art. Their milk is imbued with nutrients, purity, and wholesomeness, and cow milk clings to its status as an essential food in the diet with impressive tenacity, bolstered by a politically and economically powerful dairy lobby and mandates for it in government-funded feeding programs. Dairy products are embedded in culinary traditions and a consistent presence among foods deemed "essential' in the diet; their high value is assessed in a variety of culturally salient currencies.

Figure 1.4   The Hindu god Krishna, with a South Asian zebu cow. Indian painting, ca. 1900.

Such similarities do not mean each country's citizens are homogeneous in their praise of milk or ability to digest fresh milk in adulthood or that ideas about milk stem from the same source or are held in common. Moreover, comparison of trends over the past 40 years in each country reveals quite different milk consumption trajectories. As Figure 1.6 indicates, milk consumption has been rising rather dramatically in India since 1970 and both consumption and production overtook that in the United States in

Figure 1.5 A South Asian water buffalo *(Bubalus bubalis)*. Photo courtesy of the author.

the mid-1990s. Production of milk exceeds consumption in both countries, and has continued to escalate. However, fresh milk consumption has remained flat in the United States while continuing to rise in India. Per capita estimates provide a different perspective, as shown in Figure 1.7. Milk intake per person has been steadily declining in the United States, while in India there was a surge in intake in the 1980s, another surge around 2005, and an overall upward trend. The U.S. decline goes back to the post–World War II period, when milk consumption reached its peak. At the same time, the amount of milk the average person drinks in a year is more than twice as much in the United States as it is in India.

Exploring why two countries united by a lactophilic orientation have such different levels of intake and are experiencing quite opposing consumption trends is a major focus of this book. Not surprisingly, the underlying causes of these trends have been of interest to dairy industry officials, as well as those concerned with public health and nutrition. In India, much has been written about the dramatic rise in milk production in the late twentieth century, which came in the wake of century-long laments about India's inability to provide sufficient food, including milk, for its

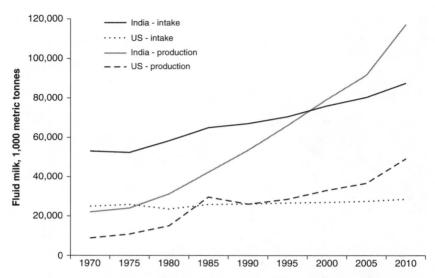

Figure 1.6   Total milk production and consumption in India and the United States, 1970–2010. Data from U.S. Department of Agriculture, Economic Research Service, 2013. ERS dairy consumption trends. http://www.ers.usda.gov /data-products/dairy-data.aspx.

growing citizenry. The production boost evident in Figure 1.6 is the result of Operation Flood, a rural development project aimed at improving the economic livelihood of the rural poor, although whether it succeeded in this regard remains an open question for skeptics. Operation Flood (also known as the "White Revolution") began in 1970 and contributed to the late twentieth-century rise in milk consumption in India, which was heralded by the dairy industry as well as public health institutions concerned about the nutritional well-being of a nation with very low animal product consumption, widespread protein and calorie insufficiency, but also rapidly Westernizing dietary patterns.

In the United States (as well as in northern European countries), a similarly spectacular rise in milk consumption began in the late nineteenth century (Atkins, 2010; Den Hartog, 2001; DuPuis, 2002). Importantly, while dairy products have featured in the United States since the colonial period, widespread fresh, fluid milk consumption is relatively recent. Prior to the mid to late 1800s, milk was more frequently processed into cheese and butter, or used in cooking. But in the current public health and nutrition literature, the focus has been on the current low level of milk consumption, which results from a 60-year-long downward trend that began after World War II. The causes of this slide are poorly understood, but much has been made of this "problem," which is seen as simultaneously contributing to the excess availability of milk (in marked contrast to India's situation), and causing lower-than-recommended calcium intakes (cf. National Institute of Child Health & Human Development, n.d.; Nicklas, 2003). There is a less visible but nonetheless vocal questioning the necessity of milk in the diet by various groups (including Physicians Committee for Responsible Medicine and People for the Ethical Treatment of Animals [PETA]), which stems from concerns about ethnic variability in lactose intolerance, dairy as a major source of saturated fat, and also from concerns about the treatment of cows for dairy production.

In both countries there is public and clinical concern about milk's digestibility. In the United States this is framed in terms of ethnic group differences in lactose intolerance—dairy advocates worry that this reduces milk purchasing and consumption while those who challenge dairy's necessity in the diet contest its presence in dietary guidelines. In India these controversies are centered on the merits of cow versus buffalo milk. Traditional Ayurvedic medicine, which is based on a humoral conceptualization

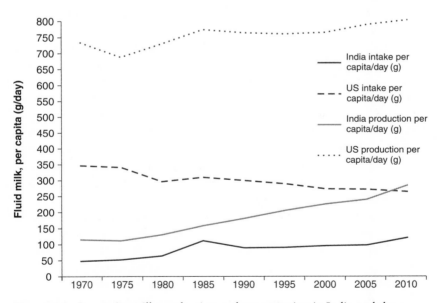

Figure 1.7   Per capita milk production and consumption in India and the United States, 1970–2010. Data from U.S. Department of Agriculture, Economic Research Service, 2013. ERS dairy consumption trends. http://www.ers.usda.gov /data-products/dairy-data.aspx.

of the body in which the body is made up of different qualities or humors, differentiates between these milk types. Individuals vary in their humoral makeup, with some having stronger "digestive fire" than others. Buffalo milk is considered harder to digest compared to cow milk, due to its higher fat content, while cow milk is seen as being more suited to those with a gentler digestive constitution. No other single food seems to prompt these anxieties in both countries (although wheat is currently a target for those increased number of individuals who suffer from gluten intolerance), where individuals appear to be attuned to the effects of different foods on their physiology.

Not surprisingly, discourse about milk also is closely related to ideas and ideals held about cows. Whether wandering the streets of Old Delhi or along the back lanes of villages grazing on garbage or grass, or, as large docile creatures in American children's storybooks, the cow is a stable feature of both countries' ecological, cultural, and mythical landscapes. But most American children see real cows—large robust beasts—only at model farms or in the distance as they drive the interstate highway system, and they are quite content with them being destined for the hamburger they eat for dinner. An Indian child, in contrast, sees marigold-garlanded yet emaciated cows routinely, popular prints featuring similarly decorated cows, some accompanying the Hindu god Krishna, and most would never contemplate eating beef.

Large bovines are "cash cows" to farmers in the United States, producing a glut of milk, but in India, laments about the cow's poor milk yield are common, and water buffalo give much more milk. On the other hand, cows are more highly prized for traction in nonmechanized agriculture (Harris, 1966). Cows rarely enter the political sphere in the United States except through their products, beef and milk, while in India, politicians from parties with different agendas (e.g., pro-Hindu or secular) wrangle over the laws to protect cows in the context of a rapidly urbanizing, technologically advanced secular state with a tradition of religious pluralism that has too often turned into religious conflict.

What is valued in milk or dairy products by consumers in India and the United States stems from their qualities as foods, about which nutritionists and public health policymakers have had much to say. Putting issues related to religious or microbiological purity aside for the moment, bovine milk has a variety of constituents that serve in its biological role as the sole food of nursing infants. These include water, fat, calcium, protein, and sugar (lactose), along with a wide array of micronutrients and biologically active substances, such as hormones and immunoglobulins, that promote growth, development, and immunological protection against local pathogens (Oftedal and Iverson, 1995; Patton, 2004). The importance of milk as a fluid should also not be underestimated because this is an infant's sole food, and dehydration from diarrheal disease is a real threat to infant health.

However, other aspects of milk's biology have captured consumers' imaginations in both India and the United States. A major one is fat. Public health campaigns against fat in food took off in the 1970s in the United States, and consumers were encouraged to consume low-fat or nonfat milk, and in general, consumers have complied (see Figure 1.8). But they have steadfastly ignored this advice in other dairy products, as cheese consumption has risen dramatically over the past 40 years. Historically, before the ubiquity of home refrigeration, production of fat-rich cheese and butter was the primary use for milk, and achieved a measure of a household's food security. In this context, fat was a primary measure of the value of milk. Milk fat in India remains a valuable commodity and is expensive. While there is no tradition of aging cheese, milk fat is clarified to make *ghee,* which can be stored indefinitely without spoilage. *Ghee* is also an inherently "pure" food and confers that quality onto those foods cooked in it. With rising rates of obesity and chronic disease related to overconsumption of calories, *ghee* (as well as whole milk) has been targeted by public health institutions in India as a food to be enjoyed only in small quantities. Indeed, what Gyorgy Scrinis (2013) has described as "nutritionist" messages about milk (i.e., those that emphasize milk's

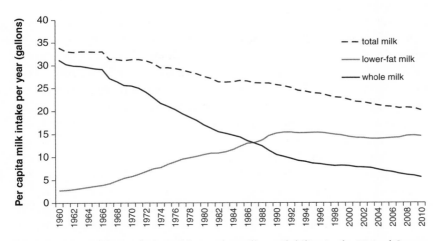

Figure 1.8   Trends in whole and lower-fat milk availability in the United States, 1960–2010. Data from U.S. Department of Agriculture, Economic Research Service, 2013. ERS dairy consumption trends. http://ers.usda.gov/data-products /food-availability-(per-capita)-data-system.aspx.

healthfulness as stemming from its nutrients as opposed to those that view milk more holistically as a food) have become commonplace, among both public health and nutrition policymakers as well as the middle class. For milk, these messages have emphasized both milk's fat (bad) and calcium (good).

Milk's association with the rapid growth exhibited by nursing infants underlies another way in which milk is highly valued and motivates parents to encourage consumption among their children. That milk might continue to enhance growth among older children—albeit in a somewhat attenuated way compared to infants—seems intuitive, although there are still only a small number of studies that have evaluated the relationship between milk consumption and growth and their results have been mixed, although overall suggestive of a modest positive effect (de Beer, 2012; Hoppe et al., 2006; Wiley, 2012; see also Chapter 5). Despite the somewhat inconsistent scientific evidence, especially that coming from well-nourished populations, there are widespread assumptions that drinking milk throughout childhood will enhance child growth, especially in height (Wiley, 2007, 2011a).

Height is associated with higher socioeconomic status in both India and the United States (Deaton, 2008; Komlos and Lauderdale, 2007) and is widely utilized in public health discourse as a convenient index of population health. A healthier population is defined as one in which children are allowed to reach their genetic potential for height, rather than being stunted through the combined effects of heavy infectious disease burdens and undernutrition (Subramanian et al., 2011). Furthermore, height is a salient measure by which potential spouses are judged. There is good evidence that height is positively correlated with marriage and reproductive success among men, at least in industrialized societies (Nettle, 2002; Pawlowski et al., 2000), but these relationships appear to be more variable across cultures (Sear and Marlowe, 2009) and for women (Nettle, 2002; Stulp et al., 2012). Some studies indicate that both sexes prefer taller-than-average spouses (Courtiol et al., 2010) and, in India, greater height is an important and overtly preferred characteristic for spouses of each sex (Smits and Monden, 2012; Wiley, 2011a). A good marriage, one that enhances a family's overall social and/or economic status, is critical in India, and appropriate spouses are chosen by parents more often than by individuals. Given the associations between height and socioeconomic status, it comes as no surprise that tallness would be a desirable attribute.

If milk is thought to enhance growth in height, its consumption may be encouraged by parents with this long-term goal in mind. Thus we would expect milk advertisers to target children and play up the benefits of milk to growth. Indeed they do. Children have been the target of milk marketing efforts since the inception of public health discourse about the value of milk in the United States, and are front and center in efforts to bolster milk consumption in India. As children are icons for the future and onto them are projected national aspirations, enhancing their size, robustness, and intellectual skills is viewed as essential. As will become evident, Indian milk ads are quite explicit about these connections. Meanwhile, in the United States, changing demography stemming from lower fertility and dramatic declines in childhood milk consumption have had the effect of decentering children from milk advertisements, while adults—including older adults—are more frequently featured. Advertisements aside, dietary guidelines in both countries strongly recommend milk, especially for children, and seek to establish a normative basis for milk consumption. This notion that children in particular "should" drink milk is well established in the United States; its acceptance across the broad and diverse populace of India is uncertain, although there is evidence for a century-long appreciation of the benefits of milk to growth.

What is relevant to both childhood milk consumption patterns and their links (desired or manifest) to growth in height is the form in which milk is consumed. "Drink your milk" is a common admonishment to American children by their concerned parents, but "eat your cheese/yogurt" doesn't have the same moral valence. Does this play out in the same way in India? How is milk distinguished from dairy products like yogurt or cheese? Is it related to milk's fluid state, or its closer resemblance to the raw material, destined for the hungry suckling infant? Does milk have a unique status among drinks, while its status as a "solid" food has a larger field of competitors? Is cow, buffalo, or goat milk more likely to promote growth?

The history of milk's normative status remains obscure in both contexts, as milk has more often been converted to fermented or precipitated foods. Routine fresh milk consumption and the cultural ideal that fresh milk *should* be consumed dates only from the late nineteenth century in Europe and the United States, and amidst urbanization, industrialization, and various scientific innovations (e.g., recognition of disease-causing bacteria in milk, pasteurization, and refrigeration). While milk consumption may have deeper roots in India, milk's normative status as a substance children "should" drink requires further investigation. How can we understand why an economic analyst in India would state, "People should want to buy milk and milk products, and have the ability and willingness to pay for them" (Saxena, 1996, 1)?

Outline

I start this analysis with historical background. The history of milk consumption in the United States is the subject of Chapter 2, and it requires a brief review of its roots in Europe. The rise of milk consumption in the late nineteenth century is already well documented (DuPuis, 2002), but its earlier—and later, and less celebratory—history has not been considered in detail. A parallel history of milk in India has never been attempted, and Chapter 3 provides an overview of what is known from prehistory and history, and outlines contemporary trends in consumption of milk. Importantly, these chapters are not intended to be definitive histories of milk in each place; my concern is more with what is known about milk consumption in the past and the various social institutions that contributed not only to milk intake but also to establishing normative views about milk in the diet. I am particularly indebted to historians such as Om Prakash (1987) and K. T. Achaya (1994b), who have combed the primary sources for references to milk in South Asia, and E. Melanie DuPuis (2002), Harvey Levenstein (1988), and John Burnett (1999), and others for their histories of milk in the United States and the United Kingdom.

In Chapter 4 I consider the sources of milk in both India and the United States. In India these are buffalo, cows, and, in a much smaller number, goats, while in the United States the cow is the only source for the commercial market. However, over the past 10 to 15 years, new "milk alternatives" have shown up in the "dairy case," which are packaged, fortified, and marketed to be very similar to cow milk. I investigate how consumers view milk from different sources, especially in relation to health and presumed biological effects, and India, in relation to the religious and economic sentiments that distinguish cows from buffalo. The debate over the cow's sacred status and how milk consumption is related to that discussion is described.

Chapter 5 asks how nonhuman milk is understood as a food that is particularly valuable to older, post-weaning-age children. The imagined and/or documented relationships between milk consumption and child growth are considered in light of the cultural valuation of these growth outcomes, especially in height. The main themes of the book are then woven together in Chapter 6.

2

A Brief Social History of Milk Consumption in the United States

The historical roots of milk drinking in India and the United States are thousands of years apart. In India, culinary and subsistence practices around milk date from the early Indus Valley civilization (3000–1500 BCE), while in the United States, these were introduced in the seventeenth century during early European colonization efforts. There was no indigenous dairy tradition, as Native American societies kept no large domesticated mammals. The countries do share a legacy of colonialism by Great Britain, although formal control by the British was later in India (1858–1947) than it was in the United States, and actual colonization by peoples from the United Kingdom was much more extensive in the United States. The consequences of these relationships with Great Britain (and other European countries such as France, The Netherlands, and, in India, Portugal) for milk consumption were quite different. The nineteenth century witnessed dramatic increases in milk consumption in the United States and Great Britain, while efforts to bolster milk intake were not really visible in India until the post-Independence period. India already had a longstanding dairy tradition; British colonialism's imprint on it is more subtle.

In this chapter I briefly review the history of milk in the United States and how it came to have its current cultural status and meanings. To understand historical trajectories in the United States it first is necessary to consider dairy usage traditions in northern Europe, as they were the basis for the establishment of such practices in the United States. Milk was probably not a major beverage until the late nineteenth century, when urban demand and technological innovations made a commercial milk supply possible. Additionally, the introduction of tea, coffee, and chocolate, all bitter products of European colonial activities, provided new ways of drinking milk. A continued upswing in milk consumption in the first half of the twentieth century was accompanied by a rush of new knowledge about nutrition, and milk became identified as a nutrient-rich food very early on, which in turn became the rationale for recommending its inclusion in the diet. Efficiencies in production and governmental support of the dairy industry provided an abundant milk supply, and large, highly productive breeds of dairy cows became the sole source for milk.

Milk consumption peaked during World War II in the United States, but has declined steadily to levels well below those of the early 1900s, despite quite popular—even iconic—advertising campaigns designed to promote milk drinking, such as the "got milk?" advertisements that have featured a wide range of celebrities. The causes of that decline are not well understood, but I argue that they are related to the meanings for milk cultivated in the early part of the century (Wiley, 2011a) and the proliferation of alternative beverages, especially those deemed suitable for children and adolescents. The historical arc of other dairy product consumption, most especially cheese, has a different pattern entirely, but the fate of these other dairy products has been bound up with trends in fluid milk consumption along with dietary guidelines that have discouraged fat consumption beginning in the 1970s. Each of these trends offers a point of intersection with India's wide-ranging experiences with milk, but also represents a continuum with European dairying traditions that go back several thousand years.

European Background: Traditional Dairy Products

Dairy production and consumption are woven into both the biological and culinary history of Europe, and date to shortly after domestication of cattle there (approx. 6,000 BCE). As noted in Chapter 1, this deep history is associated with the early spread among humans in that region of the gene that keeps lactase production on throughout life. What is less certain is the extent to which fresh milk—as opposed to cultured or separated dairy products (i.e., yogurt, cheese, butter, or whey)—was a commonly consumed beverage. Julius Caesar commented that inhabitants of the British Isles lived on milk and meat, while Strabo, the Greek geographer and historian, described the "men of Britain" in his famous *Geography* as "simple and barbarous; insomuch that some of them, though possessing plenty of milk, have not skill enough to make cheese, and are totally unacquainted with horticulture and other matters of husbandry" (quoted in Lee, 1900). Pliny the Elder also weighed in at the time, highlighting the importance of butter making among the Celts (Wilson, 1991).

While these southern Europeans expressed their observations about the dairy cultures of the north with confidence, clearly they differed in their assessments of how dairy was incorporated in the diet. This issue is important in two respects. First, the form in which milk was consumed is significant in relation to why the gene for lactase persistence spread so rapidly in northern European populations but is less common in southern European groups. Fresh milk is the sole source of lactose, as lactose is converted into lactic acid by lactophilic bacteria, or removed in the production of cheese, although it would have remained in the whey, while butter is mostly fat and has little lactose in it. Second, whether milk featured as a beverage or as a food tells us something about the role of milk in the diet. As historian of British drinking habits John Burnett notes, "milk was as much a food as drink" (Burnett, 1999, 3). The food-beverage distinction might seem a bit arbitrary; even today milk in its fluid form is frequently consumed in a "solid" context on hot or cold cereal, and yogurt might be consumed in a bowl or in a more dilute form in a glass, as is common in India. But the distinction is important insofar as it tells us whether milk was part of a meal and hence comparable to other foods, or was an add-on beverage, and as such more of a supplement to the diet. Because milk is the only "food" for nursing babies and it has a fairly high percentage of solids suspended in a liquid medium, it straddles the food and drink categories.

There is reason to believe that milk was not widely consumed fresh, but rather that its products—especially cheese, butter, and various fermented forms—predominated in the diet up through the mid- to late nineteenth century (Fenton, 1992; van Winter, 1992; see contributions in Lysaght, 1992). The lack of cold storage, especially during the summer months of high milk production, prevented widespread distribution beyond the household, or consumption long after the actual time of milking. Cheese and butter could be stored and transported without rapid spoilage from bacteria or mold, although butter can become rancid from fat oxidation. Milk was thus more often consumed as a food than a drink, and was highly valued as a source of butterfat. The liquid portion—whey and buttermilk—was used as the base for the ubiquitous grain-based porridges, turned into softer cheeses, fed to animals, or drunk in a lightly fermented state. These products would have been more readily available to rural consumers, whose ownership of a small landholding and a few animals formed the basis of their livelihood (Wilson, 1991).

Meat was more commonly preferred among the wealthy while the "white meats" (cheese, eggs) formed the basis of the commoner diet, thereby linking dairy product consumption to the poor. The value of milk products in their diets left little plain milk for use as a beverage—the cream was skimmed off for butter; the solids separated from the whey for cheese; the liquid remnants of butter making turned into buttermilk. Any of these could be mixed with honey or herbs, cereals, or meat or eggs and baked to form sweet or savory custards and pottages (Wilson, 1991). Alternatives to animal milk were also well known, primarily in southern Europe, where almond milk was used during the medieval period on fast days when animal products were forbidden (Adamson, 2004). Almond milk was also made into cheese.

The taxonomy of cheeses produced and consumed reflected their aging, source, and amount of whey retained—skimmed milk or full fat, new "green" cheese made from curd, soft cheeses matured for a short period, and then harder cheeses that kept for long periods. The latter, especially those made from skimmed milk and eaten on coarse bread, most likely formed the basis of the diet of workers on manor lands, while the full-fat versions went to the lord of the manor's family (Wilson, 1991). Butter too was widely consumed in this way, principally in Flanders and the Netherlands, and considered to be exceptionally good for growing children and the elderly, but "very unwholesome betwixt these two ages" (Wilson, 1991, 163). Butter was either salted or clarified for storage.

While the diversity of dairy products and their culinary uses were remarkable, the drinking of fresh milk was limited, except perhaps in the very northern regions. In Scotland it is reported that milk or whey was still widely drunk by the end of the eighteenth century, but whey and buttermilk had become commercial products as well and were sold in towns and cities (Lysaght, 1992). Indeed, in the seventeenth century there were fashionable "whey-houses" in cosmopolitan cities such as London (Burnett, 1999). Whey was the appropriate "milk beverage" for adults; whole milk was considered a good food for young children, invalids, and the elderly but not healthy adults (Burnett, 1999).

Whey and buttermilk—the by-products of cheese and butter making—were the mainstay of the nonalcoholic end of the beverage spectrum. These cultured liquids were part of a larger complex of fermented beverages including beer, wine, and others such as cider, perry, and mead, which were the most common drinks throughout Europe for all age groups until tea and coffee entered the picture on a large scale in the eighteenth century (Burnett, 1999; Wilson, 1991).

Material culture confirms the widespread consumption of fermented drinks, with mugs, steins, flagons, and a variety of wine glasses predominating. Milk glasses or jugs are notably rare. Furthermore, artistic representations of milk drinking are scarce in European painting. Cows, goats,

Figure 2.1 The Milkmaid. Johannes Vermeer, ca. 1658.

milkmaids, and butter churning are popular subjects, but milk is not featured in still-life paintings or those of domestic meal scenes (cf. Riley, 2004). Cheese, on the other hand, is commonly represented. Not surprisingly, milk products are most frequently found in Dutch paintings. For example, *The Milkmaid* by Johannes Vermeer (ca. 1658) shows a young woman pouring milk into a pottery crock resting on a table with pieces of torn bread (Figure 2.1). Art historians have suggested that this indicates preparation of bread pudding or a simpler dish of bread softened in milk, which was often fed to young children (Rand, 1998). Importantly, the milkmaid is not pouring milk into a glass and preparing to serve it to drink.

During the seventeenth century, as land was turned over to grain production, and common areas used for grazing were enclosed, rural farmers and workers could no longer afford to keep cows. The butter and cheese and their by-products that were produced on large estates went to markets as a commercial trade in dairy products expanded (Den Hartog, 1992; Wilson, 1991). Dairy products became more expensive and less accessible to the rural poor, and demand by growing urban populations rose. Buttermilk, which was much cheaper than whole milk or cheese, was consumed in urban areas in the late eighteenth century (Burnett, 1999). With the growth of towns and cities, water was no longer viable as a drink, not only because of contamination but because of its reputation within the humoral health framework (see Chapter 4) as "cold, slow, and slack of digestion" (Wilson, 1991, 383). Fresh milk would have similarly been a vector for pathogens, both from the source (unclean cows) and spoilage after milking, and hence beverages made from fermented grains or fruits were the primary drinks of urban dwellers.

To ease access to milk and milk products, cows were often kept in towns, either in dairy sheds where they subsisted on a variety of low-quality fodder, or left to graze in parks, where their milk could be purchased directly. By the early to mid-nineteenth century in Great Britain, household budgets indicated that fluid milk intake was very low throughout the country, averaging well under a half cup (~100g). There is some indication that it was considered important for children to have milk, and that demand for milk for them was relatively constant, such that poor families spent a larger portion of their income on milk than those with higher incomes (Burnett, 1999).

Tea and Coffee Culture: A New Role for Milk

The major challenge to the fermented beverage complex that also impacted milk consumption was the importation of coffee, tea, and chocolate in the seventeenth century. These drinks—especially coffee and tea—had the advantage of being mild stimulants, an effective antidote to the intoxication of beer or wine. Moreover, each would be turned into drinks whose palatability was enhanced by the addition of fresh milk (and sugar). Coffee and chocolate remained out of reach for most of the populace until the twentieth century, while tea was cheaper and gradually replaced beer or wine as a morning drink, particularly in Britain.

Tea was introduced into the British Isles in the late seventeenth century, an event usually attributed to Catherine of Braganza, the Portuguese wife of King Charles II, who brought the island of Bombay with her as part of her dowry. It was originally drunk in the Chinese style, green, quite weak, and without milk, but sometimes with lemon, sugar (from the British Caribbean colonial holdings), or spice (Paul, 2001). Its expense initially limited its consumption to the wealthier classes and to medicinal uses. Black fermented tea from India ultimately became the British tea of choice due to its relative cheapness, once the means to growing it commercially in India (instead of relying on the Chinese) was established, and after an assertive marketing campaign by the Indian Tea Association (see Chapter 3).

Tea, coffee, and chocolate (cacao) were originally consumed hot and "black" (or in the case of Chinese tea, "green"). All three derive their distinctive bitter flavors from alkaloids, which are produced by the plants as secondary compounds. These toxic compounds serve to deter predation and parasitism. Caffeine, theobromine, and theophylline are related alkaloids found in the three substances, with caffeine dominating in coffee and tea, and theobromine and theophylline occurring in higher concentrations in cacao. Sugar offsets the taste of these bitter substances, as does milk. The practice of adding milk to tea or coffee seems to have originated in France among aristocrats in the seventeenth century. The chatty letters of the Marquise de Sevigne mention this practice, and she declared coffee sweetened with milk and sugar to be "the nicest thing in the world" (Paul, 2001; Pendergrast, 1999). This practice spread through northern Europe and Great Britain, becoming common by the eighteenth century (Moxham, 2003), and the equipage needed to serve these became elaborated, with fine china teapots, milk jugs, sugar bowls, and cups and saucers (Paul, 2001). Thus these novel commodities were married to a traditional, commonplace food source, albeit one now employed in a novel way—both fresh and in combination with luxury items from an emergent global colonial enterprise. In this form, "a commodity that was alluring because of its very distance from the familiar would be slowly transformed into the signifier of a quotidian and very English definition of civil manners, genteel taste: the penultimate icon of civilization itself. Indeed, hidden in such consummate navigations from 'strange' to 'familiar' are the histories of empire" (Chatterjee, 2001, 21).

As access to fresh milk diminished with urban migration, and dairy products were increasingly sold to the market, tea and coffee became the primary medium for milk consumption, especially for the urban poor. The British controlled both the tea trade via the East India Company and later through direct colonial rule in India, and the sugar trade through plantations in the Caribbean. Tea with milk and sugar became an essential part of the British diet, as well as of social and domestic life starting in the eighteenth century. By the late eighteenth century the basic laborer diet included bread, bacon, tea, and sugar. Only in the northern regions of Great Britain were more porridges and dairy products consumed and tea drinking took hold much later (Burnett, 1999).

Tea consumption was ubiquitous across social classes—from service with elaborated equipage and rituals among the well-to-do to a rough-and-ready warm meal replacement for the working class. Tea could be stretched by reusing leaves, or consumed without either milk or sugar if household incomes did not permit their purchase. While the tea–sugar complex has been analyzed in some detail as emblematic of the British colonial endeavor, with its attendant global network of production and consumption, and the emergence of a thoroughly commoditized diet based around imported foods (cf. Abbott, 2008; Mintz, 1985), it is worth keeping in mind that this complex had not two, but three components: tea, sugar, and *milk,* with the latter a distinctly local product with a long history of usage in culinary traditions. Both sugar and milk were readily at hand: sugar from the Caribbean colonies, and locally produced—and familiar—milk. That said, fresh milk had never been widely consumed as a beverage, and tea (or coffee in mainland Europe) became the means by which it really entered the diet as such, especially for adults, and it too was obtained through commercial sources.

The tea-milk-sugar complex proved transformational in the British diet. It provided the idealized alternative to alcohol in a rapidly industrializing society in need of sober and diligent workers. Beer had been the staple beverage at meals from breakfast to dinner for all household members, old and young. It had been viewed as nourishing and healthful, and, along with dairy products, was produced by individual households. Beer was considered a suitable substitute for milk, and in parts of Scandinavia it was mixed with milk (fresh or skimmed or soured or whey) and drunk by all ages (Bringeus, 1992). Rising prices and land enclosures reduced domestic production and consumption of both in the eighteenth century, opening the door to commercial products, most especially tea. Both tea and milk became favored by the temperance movement as replacements for alcoholic beverages, symbolic of moral rectitude and domestic respectability and suitable for all household members. Tea thus entered into "another narrative of moral discipline and national progress" (Chatterjee, 2001, 45), and milk—both on its own and in conjunction with tea—became championed for those very virtues.

Despite their apparent opposition on the moral spectrum, both tea and beer straddled distinctions we might make between foods and beverages or foods and drugs, and both provided nutrients in a warm, rejuvenating, or relaxing medium that also inspired conviviality. By the late nineteenth century "tea" had become the tag for a new light meal between lunch and a late dinner, or the more substantial main evening meal of the working class (Burnett, 1999). Ultimately, "the cup that cheers" became the quintessential English drink. It was taken up enthusiastically in the British colonies, including those in North America, Australia, New Zealand, and later in India, the source for most tea imported into Britain. In India, passion for a highly sweetened, milky, and sometimes spiced tea *(chai)* emerged in the post-Independence era. In mainland Europe, only the Dutch took to tea drinking like the British; elsewhere coffee and chocolate (for the wealthier) became the norm, and these too were often drunk with milk. Regardless, the institutionalization of fresh milk as a beverage stemmed from its association with other drinks that had pharmacological properties, and while milk became a beverage unto itself in the twentieth century, its health-promoting and moral virtues have persisted as advertising tropes into the twenty-first century.

Colonial Legacies of Milk in the United States

Milk became part of diets in North America only after European colonization. Prior to that, Native American groups lived by hunting, gathering, or cultivation of plant food crops. When northern European settlers began arriving in the seventeenth century, their domesticated plants and animals came along and thrived in the Americas (Crosby, 1986). As a result the colonists quickly were able to reestablish many of the subsistence practices and culinary traditions with which they were familiar. These included keeping cows and making extensive use of their milk, especially for cheese and butter making. In fact, as early as 1627 there were cattle producing milk at the colony in Plymouth, Massachusetts, and ceramic wares used in dairy production were among the most common household items in use at the time (Stavely and Fitzgerald, 2004).

Cheese was a mainstay of the diet in New England, and included the hard and soft cheeses familiar to the British colonists. These, along with puddings and other milk-based dishes, were central to culinary activities (Eden, 2006). Compared to the other side of the Atlantic at the time, colonists had access to more cows and dairy products, as land was ample for grazing (Shammas, 1990). As a result, cows were bred for meat production as well as milk. Consistent with their culinary heritage, the colonists made cheese and butter from their cows' milk, and primarily consumed milk in these forms. Buttermilk or whey would have been the main way in which a fluid form of milk was consumed on its own. While this set of dietary habits became entrenched in New England and spread west with the later waves of migration from other northern European countries (The Netherlands, Germany, Scandinavia), it was not elaborated in the South, where scholars have argued that the climate was not suitable for cheese making or commercial milk trade (Levenstein, 1988; Smith, 2004).

Beverages drunken prior to the Revolution were largely fermented or distilled, rum being easily available from the sugar produced in the West Indies and whisky being easily distilled from locally produced grains. Beer, wine, and cider were drunk by adults and children, and in the late eighteenth and early nineteenth centuries, alcoholic drinks were cheaper than milk and widely considered healthful and nutritious (Rorabaugh, 1976). Tea was a widely drunk "grocery drink" (Stavely and Fitzgerald, 2004) introduced by the Dutch in the 1650s, but even just prior to the Boston Tea Party in the late eighteenth century its level of per capita consumption was only about half that of Britain. This does not appear to be due to a lack of interest in tea but rather the high import duties that precipitated the dumping of tea from British ships in the various "tea parties" precluded widespread consumption. Still, it must have been sufficiently desirable that the British tax was perceived as particularly onerous. In any event, it appears that most tea coming into the region up through the late nineteenth century was actually green tea from East Asia, much of it smuggled in during the pre-Independence period (Smith, 2004). It remains unclear whether tea was routinely drunk with milk or cream. Certainly some individuals drank it this way, and tea was routinely sweetened with sugar, but the tea–milk–sugar trio never attained a solid footing among the broader populace in North America. Iced tea, a much later innovation of the late nineteenth century, was never drunk with milk, but rather with sugar and lemon, and became a drink of the South, with broadening consumption after household refrigeration became more common.

In the revolutionary spirit of the late eighteenth century, abstention from tea became symbolic of resistance to British rule, and coffee took over as the hot beverage of choice, with similar—if not more powerful—stimulant powers. Milk and sugar became its accompaniments, although coffee was also drunk plain ("black") or with butter and an egg. In contrast to Britain, coffee was cheaper than tea in North America, since it could be grown in the Caribbean and Central America and was not subject to the British import duty. Tea never regained its position as an American drink; coffee has continued to serve in this role into the twenty-first century, and has remained an important context in which adults drink milk.

Urbanization and the Rise of Fresh Milk Consumption

The most marked change in fresh milk consumption began amidst the grit and grime of nineteenth-century cities in northern Europe and North America, when fresh milk—rather than dairy products—became a routine part of urbanites' diets, especially those of their young children. The question is why urban populations in particular began to consume larger quantities of fresh milk, a form of dairy consumption that had previously been rare.

Anne Mendelson (2008) has described the "perfect storm" that produced the explosion of fresh milk drinking among urbanites. As rural migrants moved into cities in search of employment—due to rural land reforms and opportunities for urban factory jobs—they had to rely on retail suppliers for all of their food. These vendors in turn required sources for large quantities of food commodities that could be easily brought to the urban market. Proximity was especially important for fresh foods such as milk, meat, or vegetables, as prolonged time in transport would lead to spoilage. Tea and coffee consumption was rising rapidly among most socioeconomic groups, including the working class, and milk could offset some of their intrinsic bitterness. Furthermore, without the domestic technologies of rural households, knowledge, or time to process milk into traditional forms, urbanites became the ideal market for fresh milk.

Another trend that contributed to the rise of fresh milk consumption was the growing scientific recognition of the process of "putrefication," which in the case of milk led to the familiar tangy, acidic taste of fermented buttermilk or the sourness of yeast-leavened bread. While the mechanisms underlying this process were not understood at the time, it was proclaimed undesirable as a cause of disease or digestive problems (Mendelson, 2008). This meshed seamlessly with Louis Pasteur's experiments with wine fermentation ("spoilage") and then Robert Koch's germ theory of disease in the later nineteenth century, which identified microbes as the cause of many prevalent diseases such as tuberculosis. Thus authorities concerned with public health felt justified in recommending that milk was best consumed fresh, despite the longstanding culinary tradition of milk fermentation as a method that actually retarded spoilage by pathogenic microbes.

Not only did these developments in medicine and public health stimulate growth in the demand for and supply of fresh milk, but urban working-class women were entering the industrial work force. Their long factory shifts precluded nursing their infants, and a substitute for breast milk was to be found in fluid cow milk. Urban middle-class women had become isolated from rural community networks that supported breastfeeding, and cow milk became the replacement for wet nurses. This downward trend in nursing helped create the market for commercially produced cow milk and formed the basis of campaigns that highlighted milk as the ideal food/drink for infants and children (DuPuis, 2002). However, before it could be firmly established in the public's mind that children—from infants to adolescents—needed to drink cow milk, profound problems of milk safety had to be solved.

In response to the demand for milk by city residents, dairies sprang up in peri-urban areas. Cows were densely packed into stables, which were often attached to breweries and distilleries, and sustained on the fermented grain mash by-product of beer and liquor production, so-called "swill." Not surprisingly, cows housed under such conditions were more often than not diseased, and their milk rife with pathogenic bacteria. Several bacteria species thrive in milk, including *E. coli, Salmonella, Mycobacterium* (the cause of tuberculosis), *Brucella* (the cause of brucellosis or undulant fever), and *Lysteria,* among others, and these were significant causes of illness in urban milk consumers. "Swill milk" from the nineteenth-century peri-urban dairies has been referred to as "the white poison" (Atkins, 1992), as it contributed to the very high rates of infant mortality in northeastern U.S. cities, which reached 50 percent in some areas. Of course, milk was not only to blame, as its consumption occurred amidst poor sanitation, crowding, and poverty. Physicians recognized that cow-milk-fed infants had higher mortality than breastfed or wet-nurse-fed infants, but had no means of understanding the cause of this difference.

Discourse about milk in the mid to late nineteenth century, and well into the early twentieth century tended to focus on its safety, and there was very public dialogue of the dangers attendant to drinking fresh milk, especially among children. In *An Historical, Scientific, and Practical Essay on Milk as an Article of Human Sustenance* published in 1842, Robert Hartley, an evangelical social reformist who targeted his energies at alcohol consumption and the swill milk system, raged against the latter, as well as adulteration by milk vendors:

> [Adulturation] . . . is the deterioration of an indispensable article of food, which in its best condition as furnished by the slop establishments is insufficient to support life; . . . by basely counterfeiting an indispensable food, and imposing it upon the unsuspecting for that which is not what its name imports, health is deranged, lingering and distressing diseases are induced, and life itself is destroyed with impunity. . . . We see no reason why the vending of such a drugged mixture, or merely diluted with water, is not as fraudulent and more iniquitous, than to pass pewter for silver. (Hartley, 1977 [1842], 200)

While debates erupted about the virtues of dairy inspections and milk testing (certification) versus pasteurization, which were finally settled by pasteurization mandates, what went uncontested was the "natural" role of cow milk in the diet. Hartley repeatedly referred to cow milk an "essential" or "indispensable" part of human sustenance, asserting that "it is the best and most palatable aliment for the young; it is suited to nearly every variety of temperament and is adapted to the nourishment of the body in every age and condition" (Hartley, 1977 [1842], 75). Milk's virtues were unassailable; it was inherently good and pure. Those who sought to profit from its sale without regard to the consequences (both producers who kept cows under wretched conditions and fed them swill, and the middlemen who adulterated or otherwise diluted the milk before sale) were demonized (Levenstein, 2012).

The mid- to late nineteenth century was a period of sanitary reform, and improvements in the milk production system were part and parcel of this larger trend in public health. Furthermore, focusing on contaminated milk became a convenient way in which to frame entrenched problems of urban poverty, disease, and child mortality, and provided a seemingly simple means to solve the more intractable problems of urban poverty. Indeed, as will be evident in subsequent chapters, boosting milk production or consumption has continued to be posed as an "easier" solution to entrenched social or health problems in both the United States and India.

Pasteurization was not mandated by any municipality until 1908, when Chicago became the first city to require it. The first state-level law was not instituted until Michigan did so in 1947. State rather than federal laws still regulate the dairy industry, although in practice the 1924 U.S. Public Health Service's *Standard Milk Ordinance,* which outlined standards for milk processing (including pasteurization), packaging, and selling, was widely adopted. Interstate commerce in dairy products is subject to these federal requirements.

Even with pasteurization (which at the time involved heating milk to 145°F for 30 minutes [Mendelson, 2008]),[1] more efficient methods of transportation were needed to ensure a fresh and safe supply of milk. Expansion of the rail system in the 1850s facilitated this, but by no means guaranteed uncontaminated milk because even pasteurized milk requires cold storage to retard microbial growth. Pasteurization combined with refrigerated rail cars, first patented in 1867 and in wider usage in the late 1870s, ultimately enabled mass production and broad distribution of fresh milk. Indeed, refrigeration contributed to new enthusiasm for "fresh" food, especially among a growing urban populace long (both in distance and time) displaced from traditional loci of agricultural production (Freidberg,

2009). Some scholars have maintained that it was really refrigeration that contributed most to improvements in nineteenth- and twentieth-century diets and health, particularly because it allowed more convenient access to animal products such as milk (Craig et al., 2004).

Because pasteurization and canning (first used by Gail Borden) prolonged the shelf life of milk, one constraint on milk production was alleviated, but dairies also needed facilities to accommodate this larger supply, its storage and transportation. The net effect was the merging of dairies into larger corporate units, with a centralized processing facility for pasteurization and market distribution. Processing thus became separated from production, with processors buying and pooling milk from multiple dairies. In response, there were incentives for milk producers to increase the scale of production to remain competitive pricewise. Milk, long produced and processed at home, joined the ranks of other industrial, mass-produced, mass-distributed, and mass-marketed foods of the late nineteenth and early twentieth centuries. This happened more or less simultaneously in the United States and in northern European countries, and for the first time in its history, access to fresh fluid milk became ubiquitous, along with expectations for its consumption.

Expanding the Market for Milk: Selling Milk through Its Nutrients

Commercial, community, and governmental agencies became involved in the promotion of milk, and worked together in various ways to establish and then solidify milk as an essential food in the American diet. As the volume of milk production increased, distribution networks developed, and methods of preservation were more widely adopted, expanding and solidifying the market for milk became a goal of the emerging U.S. dairy industry. The growing size and economic strength of the dairy industry, now producing a surfeit of milk, led to the formation of the National Dairy Council (NDC) in Chicago in 1915. Its goals were to support research on the healthfulness of dairy products and establish educational campaigns and other kinds of promotions for milk and dairy products. The NDC established a close alliance with the U.S. Department of Agriculture (USDA), which President Lincoln had established in 1862. As the USDA goals were to both foster and promote production and consumption of U.S. agricultural commodities and formulate government dietary recommendations, NDC and government interests were aligned, and their close relationship endures to the present (Nestle, 2002). Following the groundwork laid by nineteenth-century public health promoters of milk, the development of means of ensuring relatively safe fresh milk, and with government support of the dairy industry, the NDC was poised to gain increasing authority over the nation's health, and most especially that of its children (DuPuis, 2002; Wiley, 2011c).

Milk's construction as an exceptionally valuable food was bolstered by discoveries in nutrition, which was a very young science in the early twentieth century. In 1912, Casimir Funk coined the term *vitamin*, meaning a substance that was "vital" to life. Shortly thereafter, Elmer McCollum discovered vitamin A and B (now recognized as a complex of separate water-soluble vitamins) and named vitamin C as the means of preventing scurvy. He later named vitamin D in 1922, after discovering it to be the key vitamin that prevented rickets (as it turns out, vitamin D is unlike other vitamins in that it is not necessary in the diet because humans can synthesize it in the presence of UV-B light). McCollum and others were able to demonstrate that milk was rich in a variety of vitamins, and that these seemed to have particularly potent effects on growth and health (McCollum, 1957). Through this new lens of nutrition, milk came to be viewed as a food rich in "protective" nutrients.

As Harvey Levenstein noted in his book *Revolution at the Table*, "vitamins and minerals were an advertiser's dream" (1988, 152). They allowed marketers to be specific about the qualities of their products and back up their claims by reference to scientific discoveries. Early advertisements reference milk as a source of "mineral phosphates," then protein, riboflavin, and vitamin A; later calcium, phosphorous, and vitamin D. Milk contains relatively little vitamin D, but with the discovery that it prevented rickets (a condition in which the weight-bearing bones bend due to a lack of calcium matrix) by enabling calcium uptake, fortification of milk with vitamin D began in the 1930s. Advertisements highlighting milk's calcium started in the 1930s. Collectively, the new science of nutrition, along with the government's attempts to promote more "efficient" eating during World War I, bolstered milk consumption (Levenstein, 1988). With problems of safety solved, and increasing governmental intervention in dietary guidance supported by a dairy industry of growing influence and burgeoning nutritional science, milk became a new indispensable food in the American diet. Although the trend toward greater fresh milk consumption began amidst city squalor, it came back around to rural America, including the South. Rural farm households increasingly kept dairy cows and used their milk not for cheese or butter, but as a new twentieth-century beverage.

Increased milk drinking in the early twentieth century was part of a package of changes in American food consumption that stemmed from the move to industrialized food production, marketing, and, for the first time, public health nutrition messages underwritten by both the government and groups representing agricultural interests, such as the NDC (Nestle, 2002). In addition to milk, the overall diet was made up of more fruits and vegetables, fewer starches (cereals and potatoes), and less beef (Levenstein, 1988). Marketing of processed foods escalated, much of it touting food safety and nutritional benefits. The first dietary guidelines were issued in the 1920s, with five food groups and recommendations for their consumption as a percentage of daily calories: fruits and vegetables (30%), breads (20%), protein-rich foods (milk [10%] and meat [10%]), and "other" (fatty foods [20%] and sugars [10%]) (Welsh et al., 1993). One cup of milk was noted to be equivalent to two to three servings of meat or other protein-rich foods such as dried beans or eggs. Interestingly, cheese is not mentioned as an option within the protein-rich food group— only milk is listed. Thus milk was highlighted early on as a particularly valuable food, especially for its protein content, although other nutrients came to the fore as knowledge about them grew. Thus the push for milk as a beverage was part of this initial attempt to get Americans to eat "scientifically" with nutritional goals (however understood at the time) as the guiding rationale.

Milk Consumption in the Twentieth to Twenty-First Centuries: Waxing and Waning

In the 1920s average intake of milk was about 35 gallons per year (about 1.5 cups or 0.35 liters per day). Americans were drinking milk with meals, which was a new pattern (Levenstein, 2012), and meant that milk became an "add on" to a meal. It was moved off of the plate and into a glass. Milk consumption stayed relatively constant from then until World War II, when it suddenly surged to around 45 gallons per year (almost one pint, or 0.46 liters per day). Most likely the dramatic upsurge during World War II reflects the rationing of many desirable foodstuffs such as meat, sugar, butter, and in some cases, cheese. Fresh milk was not rationed, and production of milk increased to meet escalating demand. In addition, the military sought to provide milk to domestic troops and installations (Rifkind, 2007). Nationalist rhetoric further bolstered milk production and consumption during World War II. Milk was needed to ensure the health and strength of the troops, much as it was also essential to maintain those same qualities among the stalwart citizens at home facing rationing of their food supply. Interestingly, although British troops during World War II were comforted by cups of their familiar hot tea with milk and sugar, American soldiers were suspicious of the drink and could only be induced to drink it black (Burnett, 1999; Collingham, 2006), reaffirming national distinctions through beverage consumption.

To sustain high levels of milk consumption, school milk programs, which had begun in the 1920s as private, volunteer, or municipal efforts, were formalized as part of broader school feeding initiatives in 1946, with the National School Lunch Act (Gunderson, 1971). Fluid milk must be offered as part of school lunch programs in order to receive federal reimbursement, and subsequently private daycares and schools became eligible for these milk reimbursements. School feeding programs served the dual purposes of encouraging consumption of this food deemed especially healthy, and supporting American agriculture, most importantly, the dairy industry (Wiley, 2004). Subsequently milk became part of other government-supported feeding programs such as the Special Supplemental Program for Women, Infants and Children (also known as WIC), where it is one of a short list of foods that can be purchased through a voucher system (see further discussion in Chapter 5). Thus milk is front and center in food assistance programs in the United States, having become an important—and never politically controversial—entitlement, especially for children.

Despite a plethora of support for milk in the United States, consumption has declined steadily since its peak during World War II. The end of the war brought renewed availability of prized foodstuffs and reduced needs of the military, and milk prices fell as the supply exceeded dwindling demand. Ironically, during World War II milk was joined by Coca-Cola as another drink with nationalist ambitions. Coke was recognized as so important to soldier morale that it was exempt from sugar rationing. It became an important symbol of American nationalism—a true "taste of home" (Mintz, 1996). Milk might have been what people at home and at the front "should" drink, but sweet and stimulating Coke was what they wanted to drink, a distinction that remains intact, and which may explain why U.S. soldiers disdained British milky-sweet tea. Milk never regained its pre–World War II popularity, while consumption of Coke and other soft drinks has risen substantially (Duffey and Popkin, 2007).

Based on national-level surveys of milk consumption, in 2007–2008, less than a quarter of adults age twenty and above reported drinking milk or a milk product on any given day, and average milk intake was under a half cup (120g). Most drank water (over 75%), and the majority reported drinking coffee or sodas, as shown in Figure 2.2. Most tea and coffee was drunk with milk, although one-third to one-half was drunk plain. Less than 10 percent of adults reported drinking milk at breakfast or dinner, with even lower frequencies for lunch or snacks (LaComb et al., 2011). The milk-drinking habit of the early twentieth century turned out to be short lived, as the glass was filled with other beverages.

At present, per capita milk consumption estimates based on market data show an average milk intake of about 250g (~1 cup) per day (just under 25 gallons/year), but a steady decline over the second half of the twentieth century and into the 2000s, as shown in Figure 2.3 (U.S. Department of Agriculture, Economic Research Service, 2013; see also Figure 1.7). This is about one-half the level that it was at its peak in 1945 and also well less than the amount consumed in 1909 when the data are first available. What Figure 2.3 also indicates, however, is that while milk consumption has been declining, cheese consumption has dramatically increased over the same period.

What factors are responsible for these changes? Certainly the proliferation of other drink types contributes, although the downward trend be-

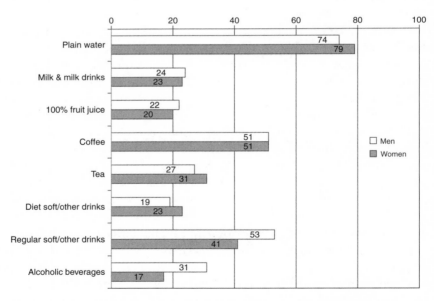

Figure 2.2   Percentage of American adults age twenty years and older reporting consumption of different beverages, NHANES 2007–2008. Redrawn using data from LaComb et al., 2011. Beverage choices of U.S. adults: What we eat in America, NHANES 2007–2008, Food Surveys Research Group, www.ars.usda.gov /SP2UserFiles/Place/12355000/pdf/DBrief/6_beverage_choices_adults_0708.pdf.

gan long before these appeared in such large numbers. Soft drinks and other "sugar-sweetened beverages" are now more widely available and contribute more calories to the diet than milk (Popkin, 2010). The sheer variety of drink options emphasizes to consumers that options other than milk exist. While milk is still a ubiquitous part of school feeding programs, schools have also signed pouring rights contracts with soft drink companies that allow them to place vending machines in schools in return for funding (Nestle, 2002). Eating outside of the home is another phenomenon likely linked to declining milk intake, as consumers—including children—are able to choose what they like and are encouraged to "indulge" in drinks like sodas. As is evident in Figure 2.4, total dairy intake from sources outside of the home was only about one-third of dairy consumption within the home in 2003–2004, and this is true for both adults and children, although children consume somewhat more dairy products outside of the home than do adults, probably due to their consumption at school. And, although milk has long traded on its nutritional value, the proliferation of health claims on beverage labels since the turn of the twenty-first century as well as the fortification of other beverages (e.g., calcium in orange juice or soy milk) has eroded some of milk's unique status. Milk must now compete for a "share of the throat," especially outside of the home (Den Hartog, 2001).

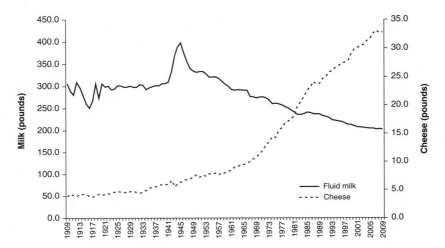

Figure 2.3 Milk and cheese available for consumption: United States 1909–2009. Based on data from http://www.ers.usda.gov/data-products/food -availability-%28per-capita%29-data-system/food-availability-documentation .aspx#dairy.

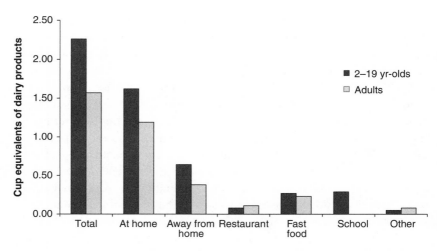

Figure 2.4   Dairy consumption within and outside of the home, NHANES 2003–2004. Based on data from http://www.ers.usda.gov/data-products/food -consumption-and-nutrient-intakes.aspx#.UYFqOsrhdPt.

Contemporary Meanings of Milk in the United States: Consumption and Production

What does milk mean in the contemporary United States? How do those meanings shape peoples' consumption practices, most especially their declining milk intake? And, how does milk's source, which in the United States is exclusively the cow, figure into these? First of all, despite declines in actual consumption, milk does retain the normative status that it gained a century ago. Milk advertisements are rife with "should" messages. Children are frequently targeted with admonitions to "drink your milk" by their caregivers. Government-sanctioned dietary recommendations such as the U.S. Dietary Guidelines have consistently featured milk since their inception in the early twentieth century, even if the "food groups" have changed their form and health concerns have shifted from undernutrition to the negative health consequences of overconsumption. Current recommendations are for three cups of low-fat milk per day, or the equivalent dairy amount (e.g., 2 cups of cottage cheese, or 1.5 cups of ice cream! [Figure 2.5]).[2] School and day care meal programs still require milk to be offered, and the government underwrites the cost of milk in these and other food assistance programs. Every refrigerator "should" have milk in it—one need only see the empty

Figure 2.5   MyPlate, the visual representation of the 2010
U.S. Dietary Guidelines.

shelves in the dairy case at a grocery store before a major storm to understand that Americans view milk as a "must-have" commodity in their
refrigerators.

The targets of marketing messages for milk have shifted, reflecting demographic changes in the United States as well as research suggesting
some relationships between milk and chronic disease. Early twentieth-
century efforts focused on children—preschool and primary-school-
age children—in an effort to expand the market for milk with a message
about milk's salutary effects on child growth and health (see Chapter 5).
These demographic groups are no longer featured in milk promotions
(Wiley, 2011a). Online milk promotion websites (such as www.whymilk
.com, sponsored by the National Fluid Milk Processor Promotion Board,
now at the URLs www.thebreakfastproject.com, or www.gotchocolate
milk.com sponsored by the Milk Processors Education Program) change
rapidly, and their messages do as well. In part, this ongoing alteration of
milk messages stems from the Food and Drug Administration and
Federal Trade Commission's regulation of health claims made for advertised products (in 2007 the FTC pulled the dairy ads claiming that dairy

consumption would contribute to weight loss). Both sites are currently highlighting milk's ability to "fuel" physical activity and enhance athletic performance. The latter targets athletic adults, and features chocolate milk's utility as a "refueling" beverage to aid in post-workout recovery. The former emphasizes breakfast and "fuel up to play 60," the current marketing slogan from the National Dairy Council in concert with the National Football League and the USDA (found at www.fueluptoplay60 .com). This links milk to current efforts to encourage physical activity (60 minutes per day) and school meals (breakfast and lunch).

Milk has long been positioned as a "problem-solving" food, perhaps more so than any other food. Can't get your child to eat? Have them drink a glass of milk and get a rich package of nutrients. Worried about osteoporosis? Milk has calcium that is important to maintaining bone density. Worried about your weight? Milk could "support healthy weight loss." The problems milk "solved" in the early twentieth century are quite different than those it attempts to solve today—although arguments supporting consumption of pure milk to offset grinding urban poverty find an echo in recent rhetoric that low milk intake contributes to higher rates of chronic disease among ethnic minorities in the United States (cf. Jarvis and Miller, 2002). In the first half of the twentieth century the problem to be solved was child growth, building stronger bodies to ensure national safety (during World War II), or reducing problems with rickets. Currently the focus is on healthy weight (by pairing it with exercise), or reductions in chronic disease risk (see especially the vitamin D claims or osteoporosis claims). Some recent examples include the 2009 "Milk is nature's wellness drink!" promotion:

> People who regularly drink milk tend to have healthier diets that are richer in nutrients. They're also more likely to be at a healthy weight. Milk drinkers may also have an edge when it comes to cardiovascular health, including blood pressure. Plus, no other beverage offers the same array of bone-building nutrients, including calcium, vitamin D, protein and phosphorus, and has hundreds of studies supporting its link to strong bones. And there's also emerging evidence that the nutrients in milk may play a role in reducing the risk of cardiovascular disease, certain cancers, type 2 diabetes and metabolic syndrome. (Milk Processors Education Program, 2009)

This promotion gave way to a 2010 "Building Strong Families" theme. In this marketing scheme, milk was no longer emphasized for middle-age women, but rather for the whole family, with benefits outlined—in general form—for each age group.

While the milk industry struggles to find a message or a new demographic to stem the ongoing slide in milk intake (as evidenced by their rapidly changing promotional schemes), it seems to be returning to its relatively secure spot in school feeding programs. But another approach is also being taken. A recent milk promotion found on www.gotmilk.com began with a "find the real milk" activity. Old-fashioned glass milk bottles of whitish fluids are lined up, and as the cursor passes over each, certain additives are highlighted in the nondairy milks: carrageenan, xanthan gum, or guar gum. When the "real milk" is selected, the label reads: "real milk comes from cows." Scrolling through the website reveals a Holstein cow wagging its tail and ringing the bell around its neck while standing on a mound of green grass in a celestial setting. Trading on purity and idyllic bucolic roots, this effort is designed to fend off competitors in the dairy case—the myriad soy, almond, coconut, hemp, and other milks that are claiming valuable grocery store space and packaged and formulated to be "just like" or "even better" than "real milk."

There is much irony in milk's current claims to purity, given a past rife with adulteration and contamination, not to mention the ongoing fortification (vitamins A, D) or removals (especially of fat) that all "milks" undergo to enhance their nutrient profile and "fit" the nutritional needs of the moment. Low-fat dairy products in particular have a number of additives to enhance their texture. Moreover, as we've seen, fresh milk consumption did not begin on the family farm, but rather amidst urban squalor, with cows penned in filthy barns rather than grazing on wholesome green grass.

Cows and Milk Marketing

Especially in relation to milk in India, the question emerges as to the significance of the cow in milk ads or in Americans' dairy consumption practices. In a 2006 exhibition featuring cows in American art ("Got Cow?"), it was noted that the cow symbolizes "a simple kind of domesticity, an earthiness . . . a desire for nature and innocence; they elicit memory, nostalgia, and reverie; they are soulful and spirited" (Bland, 2006, 4, 6). Typically a subject for pastoral landscape paintings and folk art, where cows are featured in their agricultural context, contentedly grazing or being milked by industrious milkmaids, dairy cows are usually associated with females, and are docile, yet protective, mother figures, sometimes pictured as members of the nuclear family. Currier & Ives prints from the late nineteenth century often featured such pastoral landscapes with cows (see Figure 2.6), and became ubiquitous in middle-class urban parlors, reflecting nostalgia for a more innocent rural past while confirming domestic solidity and respectability.

Figure 2.6   American Homestead in Summer. Currier & Ives, ca. 1868–1869.

The demand for cow art faded by the turn of the twentieth century. Representations of the cow persisted in regional genres, but were only revived when they gained further artistic purchase in Andy Warhol's modernist cow wallpaper, with repetitions of garish purple cow heads against a lurid yellow background. In this and other late twentieth-century representations, the cow is separated from its landscape, commoditized and commercialized (it is after all, Elsie, the cow featured on Borden tinned milk products). Cows have remained an article of kitsch in the United States, with Holstein figures made to fit into every imaginable household niche.

But have cows been used to sell milk? In the mid-nineteenth century, pristine cows and demure milkmaids were frequently featured on milk brand labels, mimicking the pastoral art of the time. As DuPuis notes, "The milkmaid, overseeing the purity of milk, therefore symbolized this nostalgia for the nurturing countryside, this idea of perfection as the antidote of the downfallen city (2002, 97). These images gave way in the early twentieth century to images of healthy, robust, white children contentedly drinking their milk (Wiley, 2011c), similar to those painted by Mary Cassatt (Figure 2.7). This marks a shift in emphasis from concerns about production to the benefits of milk consumption after pasteurization and anti-adulteration law resolved some of the problems with milk purity. The distinction between images of cows or of people drinking milk that are used in milk advertising remains, with some changes in meaning. For example, local dairies frequently use images of cows in their packaging, and the California Milk Advisory Board (CMAB), which represents "dairy families" in the state, features a number of cows in its advertisements, many of whom are depicted in the house interacting with family members. An entire section of the CMAB website (www.realcali forniamilk.com) is devoted to photographic or other artistic representations of "happy cows."[3] As depicted in milk commercials, they are selling

Figure 2.7   Child Drinking Milk. Mary Cassatt, 1868.

the virtues of rural life as well as a cozy domesticity, but they also present a plea to help save the family farm from the onslaught of industrialized agriculture (this is also true of cows used on organic milk labels) and offer a defense against claims of cow maltreatment in the effort to increase milk productivity.

The placid, portly cow would seem to be at odds with, and is rarely featured in promotions emphasizing milk's workout benefits or its relationship to chronic disease risk, although it is frequently used in nutrition education materials (produced by the National Dairy Council) aimed at young children (cf. http://school.fueluptoplay60.com/, "Thank Goodness for the Cow"). Thus how milk is advertised reflects two major contemporary concerns about food: the erosion of traditional farming and renewed interest in local "whole" foods, with family-run dairy farms being symbolic of threats to small-scale agriculture, and public health concerns that have moved from child undernutrition to chronic health conditions of the growing middle-aged and elderly populace. The new "find the real milk" promotion brings back concern with "purity" by juxtaposing model cows and their milk against industrial food additives. Just as the nineteenth-century use of the cow in milk advertisements provided an antidote to the ills of city life, the separation of urban consumers from producers, emergent industrialization, and milk contamination and adulteration, the dairy industry is reviving this theme by raising fears about scary-sounding and possibly dangerous additives in other nondairy milks that look very similar to "real" milk.

Cows in milk commercials can also mediate the realms of production and consumption. Dairy cows have long been portrayed as fertility and maternal figures, even the "foster mother of mankind" (White, 2009). They fill that role as producers of the milk consumed in some form or another by babies and children in societies with dairy traditions. Large and solid, dependable, seemingly uncomplainingly docile, yet fierce protectors of their young, and producers of a steady stream of wholesome milk, they easily conjure up maternal sentiment. More often, images of cows are replaced by human mothers dispensing milk into a glass for their children, modeling good milk-drinking behavior, and gently chiding children to drink their milk in the service of their well-being. Mothers are the protectors of child health through their role as food, especially milk, providers.

Current Trends in Dairy Consumption: Back to the Future

Despite the efforts of milk producers and processors to entice contemporary Americans back to their "tradition" of milk drinking, as Figure 2.3

clearly shows, Americans drink less milk each year. But it is also evident
that there has been renewed enthusiasm for cheese. Annual per capita
cheese consumption has risen by a factor of eight (4 to 33 lbs.; 1.8 to 15 kg)
since the early twentieth century. Within the past 40 years, while milk
consumption has declined by 25 percent, cheese consumption has almost
doubled. It is rather ironic that cheese, which is high in fat, is enjoying
renewed popularity given that consumption of whole milk has been giving way to lower-fat milks across all age groups (Figure 1.8), and this has
been celebrated in public health nutrition as one of the few examples of
a successful diet-education campaign to encourage healthier eating (read:
lower fat consumption) at the national level (Nestle, 1998). This trend,
combined with the rise of yogurt consumption, has kept overall dairy
intake relatively steady.

Yogurt intake also rose in the 1970s, and has had a more spectacular
increase, particularly since 2000, when per capita intake more than doubled, from 6.5 to 13.5 lbs. per year in 2010 (Chandan, 2013). Yogurt
had never enjoyed the same level of consumption in the United States as
in Europe, especially in the Mediterranean, but current interest in "the
Mediterranean diet" and "Greek-style" yogurt as its standard bearer, as
well as interest in probiotics (helpful bacteria in the colon) and their benefits to "digestive health" for a population apparently suffering from con-
stipation contributes to its growing presence among dairy products. While
cheese is consumed across the age spectrum, probiotics are marketed primarily to adults (and among them, women). Heavily sweetened yogurt is
readily available and marketed to children. It is worth remembering too
that a by-product of fermentation or cheese making is a reduction in lactose, and while cheese and yogurt are not specifically marketed for this
characteristic, it may facilitate their consumption among diverse minority
groups with high rates of lactase impersistence.

Both trends, particularly increased cheese consumption, can be understood as a return to earlier traditional patterns of dairy usage. While earlier
forms of milk production and preservation made ample use of fermentation by bacteria, these had been denigrated during the heyday of the sanitation movement (which was undergirded by the germ theory of disease).
However, they have been reinvigorated, and especially in the case of yogurt, in the form of high-priced value-added dairy products, rather than
through home production. Even if not produced within the home, yogurt
and milk are still largely consumed there, while cheese offers a means by
which dairy can be consumed outside of the home, which is where an
increasing proportion of meals occur. Thus while Americans are no longer consuming milk in a glass (despite the MyPlate recommendation,

Figure 2.5), they are increasingly eating it on their plate, in a variety of cheese-laden dishes, most especially pizza, as about one-third of all cheese consumed is Italian-style mozzarella. When Americans do drink milk, it is likely to be heavily sweetened, like most other beverages in the American diet (Popkin and Nielsen, 2003).

Although nutritionists and the dairy industry alike decry the decline in milk consumption among Americans, it may be that widespread consumption of fresh milk was more of a historical interlude rather than any representation of the "normal" American diet, spurred on by the novelty of easily available and safe milk, the marketing of milk as a nutrient-rich food during a time when nutrients were just being identified, and support by the government through dietary recommendations and feeding programs. Surely fresh milk will not disappear from the American diet, and it remains a normative drink for children, but its predominance as a drink of the early and mid-twentieth century has come to an end. Nonetheless, dairy products remain well-established in the American diet, even if Americans are not consuming them according to the recommendations of dietary policymakers.

3

A History of Milk in India

The relatively recent history of milk usage in the United States contrasts with that of India, which had indigenous bovines suitable for domestication and early experience making and using dairy products. As a result, dairy culture was deeply embedded in Indian society and culinary traditions, but despite its legacy and the more recent efforts of Operation Flood to boost production, twentieth-century trends indicate very low per capita milk and dairy product production and consumption levels compared to the United States (see Figure 1.7). However, India has been experiencing an upswing in intake over the past few decades, while milk consumption has been declining in the United States. Greater intake is largely an urban, middle-class trend, yet India's population is mostly rural and poor. Differences in milk consumption mirror other socioeconomic disparities in India.

In addition to comparisons of consumption trends and the social contexts in which these are occurring, central questions and points of comparison with the United States are how milk has been used (i.e., has it been drunk fresh as a beverage or converted into other forms through fermentation or separation), how experience with British colonialism impacted India's dairy culture, how the colonial tea enterprise impacted milk consumption, and desires for, and anxieties about, the milk supply in relation to public health. Equally important is how milk consumption is related to evolving cultural and religious practices surrounding the sanctity of the cow. Milk consumption over the twentieth century and now into the twenty-first century in India has been linked to political and religious agendas, beliefs about milk, and increasing authority of nutritional science to define milk and prescribes its usage.

Early History of Dairying in India

In South Asia (Figure 3.1), both zebu cattle *(Bos indicus)* and water buffalo *(Bubalus bubalis)* are indigenous domesticates. Archaeological and genetic evidence indicates that zebu cattle were domesticated from a South Asian wild progenitor *(Bos primigenius)* around 8,000 years ago in Northwest India (Chen et al., 2010; Fuller, 2006; Meadow, 1996). Zebu are characterized by a distinct humped fat deposit between the shoulders

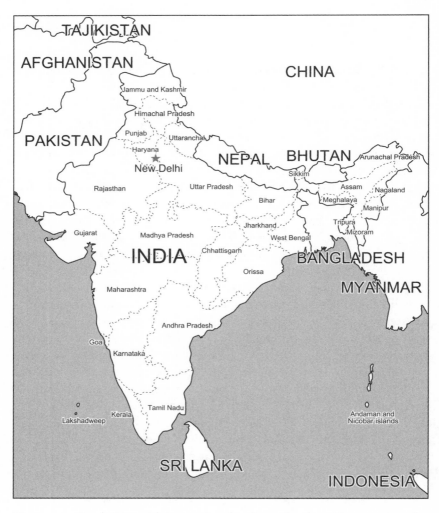

Figure 3.1   South Asia, with international and state borders. Map courtesy of http://www.your-vector-maps.com/.

Figure 3.2   Zebu cattle with their distinct humps, with a goat at a farm in
Maharashtra. Photo courtesy of the author.

(see Figure 3.2) and are well-adapted to the warm climate of South Asia.
There were likely other later sites of domestication in the Indo-Gangetic
Plain and in South India, as evidenced by large mounds of cattle dung
ash (Chen et al., 2010; Fuller, 2006; see also Figure 3.3). Zebu cattle were
fully domesticated by the time of the early Indus Valley civilization (ca.
3300–1300 BCE; also known as Harappan culture), as shown by osteological remains and ceramics indicating their use for milk as well as traction (Fuller, 2006). Harappan seals show the characteristic hump-backed
and dew-lapped zebu cattle, as well as western humpless taurine cattle
(Achaya, 1994b).
   The water buffalo is similarly indigenous to South Asia, and while the
cow receives more attention as an animal of great cultural significance,
buffalo produce much more milk, and milk with more butterfat. Like
cattle, wild populations were present in South Asia long before domestication, and by the time of Harappan culture, they were being herded and
used for plowing, milk, and possibly meat. Goats and sheep are also found
in Harappan sites; whether these were milked is not clear (Fuller, 2006).
Regardless, there is ample evidence of extensive use of domesticated
cattle, as well as water buffalo, goats, and sheep in agricultural and urban

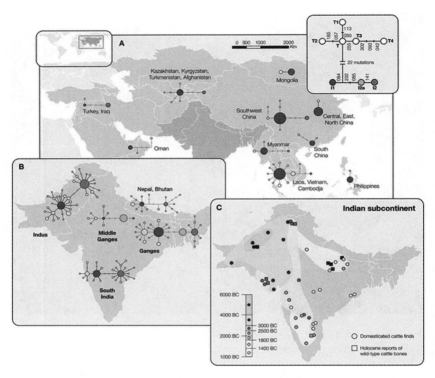

Figure 3.3 Geographic distribution of between mtDNA genetic patterns across Asia and the archaeological signs of spread of cattle pastoralism within the Indian subcontinent. (A) Median-reduced networks constructed for zebu haplotypes across Asia; (B) Indian subcontinent showing median reduced networks for each potential domestication center (Indus, Ganges, and South India); (C) Indian subcontinent indicating the spread of cattle across time based on archaeological data. Circles represent sites containing domesticated zebu cattle faunal remains, and squares represent reports of Holocene wild-type cattle bones. © 2010 Chen, S., et al. Published by Oxford University Press on behalf of the Society for Molecular Biology and Evolution. All rights reserved.

sites in the Indus Valley region by the end of the third millennium BCE (Meadow, 1996).

While the Harappan culture developed urban centers that were supported by rural agricultural production, pastoralist populations were also present in other parts of the subcontinent during the same time. As the Harappan culture began to fragment during the second millennium BCE, pastoralists became more prevalent and populations began to disperse east and south. Simultaneously there was migration of Indo-European-speaking herding populations from the northwest, who appear to have contributed to the south and east movement of the Dravidian-language-speaking occupants of the Indus valley (Thapar, 2003). Genetic evidence supports a later arrival of Indo-European populations, and substantial admixture with indigenous Dravidian populations, who are now most numerous in southern India (Basu et al., 2003).

The Indo-European migrants were agropastoralists, cultivating grains such as barley and consuming the products of their animals as they settled in northwestern India and migrated throughout the Indo-Gangetic Plain in search of grazing pastures for their animals. Oxen were used for plowing while cows were used for milk, and there is evidence that cattle were also eaten (Jha, 2002). They established a social organization based on four ranked castes (Brahmins [scholars and priests], Kshatriyas [warriors], Vaishyas [herders, farmers, tradesmen], and Shudras [laborers, service providers]) and produced an oral literature called the Vedas (produced between ~1500 BCE and 500 BCE). The oldest Veda is the Rigveda, thought to have been composed around 1700 BCE. It contains over 700 references to cows, who symbolize endless bounty or blessings. Cows were *kamadugha*, meaning "milking desires," "yielding objects of desire like milk" or "yielding what one wishes." This Sanskrit literature was eventually transcribed and forms the foundational texts of Hinduism. The Vedas are a rich source of evidence for normative views of food consumption; they describe what should be eaten, or perhaps what was eaten by Brahmins, but they should not be interpreted as reflective of actual food consumption of the population.

In the Vedas there are numerous allusions to cow milk, as well as some mention of buffalo and goat milk (Prakash, 1987). These were consumed in several forms: fluid milk, butter, cream, buttermilk, and yogurt. Fluid milk was drunk fresh or boiled, and was frequently mixed with other ingredients (spices, or the mysterious intoxicating substance *soma*) and used in the more quotidian form of porridges. As Om Prakash writes in his history of food in ancient India, "Milk formed one of the principal ingredients of the food of Vedic Indians. Generally boiled cow milk was taken. It was used in preparing a mess with grains and a gruel with poached barley flour," (Prakash, 1987, 12–13). While this does not sound particularly appetizing, it is very similar to the milk and grain-based porridges that sustained northern Europeans well into the twentieth century.

Along with milk, yogurt (usually called curd in English in India) is one of the oldest of Indian foods—in ancient India there were at least four different names for it. Curd is formed from heated milk mixed with already soured (fermented) milk left in a warm moist place for the curds to "set." It had seemingly infinite uses: it was drunk on its own or diluted with boiled milk, eaten as a meal by itself or with grains, or mixed with sweet flavoring to make confections. Butter was made from the churning of curds and produced buttermilk as a by-product, which was also drunk. The butter was then heated and the remaining solids removed, forming *ghee*. In this form *ghee* is not vulnerable to spoilage, and would have preserved milk fat for consumption during the seasons when dairy animals were not giving milk. A mixture of curds and butter was known as *prsadajya,* a common offering to the gods (Prakash, 1987).

Two kinds of cheese were described: a fresh cheese—possibly simply curds that have been allowed to drain and/or are pressed—and a harder, ripened form, but these were neither elaborated nor consumed as widely as curds, milk, and *ghee*. Indeed there may have been some prohibitions on "cutting curds" with acid or with rennet to make cheese. The Portuguese are generally credited with using such techniques to make contemporary cheeses like *paneer* (a fresh cheese separated with a strong acid) and its derivatives, many of which are used as the basis of sweets, especially in Bengal (Banerji, 2006). Overall, milk was cultured to produce several products, all of which were incorporated into South Asian diets, and butterfat, clarified into *ghee*, was the most highly valued of all of these.

In the Rigveda, much is written about the mysterious substance, *soma,* which was often drunk mixed with curds, *ghee,* or milk. *Soma* was a plant whose stalk was pressed to form an ambrosia that ensured immortality. As such, it was considered the drink of the gods, much like cacao was blended with various seasonings to make a spicy, frothy drink the Aztecs called chocolatl, which was reserved for consumption by elites (Coe et al., 1996). This pure, heavenly beverage was said to inspire courage, confidence, and alertness. *Soma* may have been a variety of *Ephedra,* a plant known for its stimulant properties. While consumption of intoxicants was forbidden among Brahmins, *soma* was considered their beverage, while curds were for Vaishyas, water for Sudras, and root extracts were for Ksatriyas (Prakash, 1987). Importantly, milk and curds produced from *soma* were similarly considered "immortal nectar," as substances derived from the sacred cow. Their addition to *soma* likely added to the allure of this drink.

As populations from the Vedic period grew and expanded further into the Indian subcontinent, they formed more permanent agricultural settlements and city-states that coalesced into the Mauryan Empire (322–185 BCE), which stretched across the subcontinent. With a decline in pastoralism, most of the population by this time was engaged in mixed agriculture, growing grains and pulses, and keeping cows, buffalo, goats, and sheep in village settlements. While there was a greater emphasis on grains and pulses, milk and milk products retained a central position in the diet, at least from what we know of life described in the Brahamanas, a series of commentaries on the Vedic literature. Milk and its products were one of the main food groups recognized, along with cereals, pulses, vegetables, fruit, spices, animal meats, and alcoholic beverages. This may represent one of the earliest forms of dietary guidelines, and current Indian dietary guidance looks very similar (albeit with vegetarian alternatives and limits on alcohol consumption). Milk production and processing became a commercial endeavor as well as a domestic one, with specialized occupations for milking and processing of milk. Thus milk was well established early on in Indian life as a commodity with market value as well as a food produced and consumed within the household.

The value of cows was not limited to milk, and their other uses may have been more critical to the livelihood of their owners. Bulls provided traction and their dung was a source of fertilizer and fuel. Furthermore, cows and buffalo were eaten. Prohibitions on the consumption of cows gradually emerged and spread over centuries (Jha, 2002). In the Vedic period cows were sacrificed to the gods and to honor guests, and later became restricted in Brahmins' diets. With the advent of Buddhism (600–400 BCE) and its emphasis on *ahimsa* (doing no harm), killing cows for consumption or sacrifice declined, although there was no ban on their consumption after a natural death. There was some conflict in early Buddhism over whether milk drinking itself constituted harm, as the calf was deprived of its nourishment, but this view—perhaps not surprising in this cultural and culinary context—was ultimately discarded.

The Laws of Manu, written in the early years of the Common Era (CE), and other Dharmasutra texts prohibit Brahmins to consume cows, although they are not singled out as a separate class of forbidden animals, and in the case of starvation, consumption of just about anything was allowed. Overall there appears to be some ambivalence about cow consumption, but this concern extended to meat consumption or animal slaughter in general (Olivelle, 2002). As Wendy O'Flaherty Doniger has argued, models for ideal Hindu devotee practices underwent a transition from animal sacrifice and consumption of bulls to milking of cows as models for the ideal devotee (1980). As we'll see later, laws governing the killing of cows and movements to protect cows became part of various political agendas in India in the twentieth century.

Milk and *ghee* came to have both social and religious significance, although importantly, milk was acceptable only from cows, buffalos, and goats, and with some additional restrictions (not from the 10 days after birth, from those who had twin offspring, or who are pregnant or who have lost their calf [from the Gautama Dharmasutra]). Milk and *ghee* are considered "pure" substances, rendering any food cooked in them pure and invulnerable to pollution. Milk, *ghee,* and curd constitute three of the five sacred products of the cow *(panacagavya),* the other two being urine and dung. Codes for the sharing of food of different kinds are fundamental to Hindu social interactions, and there are complex rules regarding what kinds of people can share what kinds of foods. In general, foods cooked in milk or *ghee* are protective (there seems to be some debate over the status of curd [Achaya, 1998]), and may be shared across caste lines more easily than those cooked in vegetable oil or water, which are vulnerable to the pollution associated with lower-caste individuals. At the same time, the gods prefer cooked foods, and all Vedic sacrifices required cooking, which raised a potential problem for the use of milk. According to Charles Malamoud, "milk and consequently all milk products and derivate substances in which milk are blended are, as far as Indian physiology is concerned, cooked in advance" (Malamoud, 1998, 38). This was because milk was considered to be the sperm of Agni, the Vedic god of fire, and all that comes from Agni is by its nature already cooked.

By the beginning of the Common Era there is unambiguous evidence of milk being widely consumed as a beverage—whether fresh, boiled, or as buttermilk, whey, or with curd or spices of various kinds added. It does seem to have been boiled as a routine practice, and the idea of a cold glass of milk would have been unheard of. As Om Prakash notes, "milk and its other products such as buttermilk were generally used as beverages" and, writing about the period from 300 to 750 CE, "milk continued to be the favorite beverage in India" (1987, 187–188). In addition to milk as a drink, fruit syrups and spiced drinks were common. It is also important to note that water was widely drunk. While it may seem surprising to that this would even be mentioned, it has long been argued in Europe that alcohol was consumed in lieu of water because water would likely have been contaminated (cf. Burnett, 1999; Wilson, 1991). Indeed, in contemporary India, water is often a source of gastrointestinal pathogens, including cholera epidemics. But water was celebrated as a drink in the past. In the Mahabaratha, water is held up as "the best of beverages." Up at least until 1200 CE there are references to water from rain, streams, lakes, wells, or springs as a nectar, though in later years it is also recognized as a potential poison (Prakash, 1987). Even by the time of British colonization, Chaplain Edward Terry remarked, "That most ancient and innocent Drink of the World, Water, is the common drink of East-India;

it is far more pleasant and sweet than our water" (cited in Collingham 2006, 189).

In the early medieval period (1000–1500 CE), descriptions of royal feasts among the regional elites often featured milk or curds toward the end of the meal, with milk consumed in a sweetened form and *ghee* used liberally in food preparation. Of course, we do not have any solid sense of what kinds or how much milk or other dairy products most ordinary individuals would have been consuming. But foreign travelers to India did not tend to comment on extensive milk usage, perhaps being more captivated by the exotic spices and fruits available in the region. Only the Chinese Buddhist monk Xuan Zang, who traveled extensively in north India in the seventh century noted that milk butter and cream were among the "most usual foods" (Prakash, 1987, 148). Almost five hundred years later the Italian Catholic missionary John of Montecorvino wrote "The people of India were scrupulously clean, feeding on milk and rice" (Prakash, 1987, 163), while others noticed the abundance of domestic cattle. There is mention of water as the common drink and that the poor subsisted on rice with pepper and butter, *chappatis* (flatbreads made from wheat), or pulses with rice and butter, suggesting greater usage of *ghee* than fluid milk or curds.

By the time of more extensive European experience in South Asia, pastoralism had pretty much died out, but dairy animals were essential to household livelihoods and dairy products were clearly part of the diet. William Macintosh, writing of his travels in India, noted only that, "The diet of the Hindoos is simple, consisting chiefly of rice, *ghee* (a kind of imperfect butter), and oriental spices of different kinds" (Macintosh, 1782, 329). Slightly later, Francis Buchanan Hamilton, surveying southern India for the Governor General of India, remarked on household milking practices, noting the greater productivity of buffalo and that sufficient cows were kept by most farmers only to supply their families with milk. Milk could be purchased from the market, or from small stalls serving travelers. His Bengali companion reportedly expressed much delight in the abundance, superiority, and relative cheapness of milk and curds in what is now the state of Karnataka (Hamilton, 1807).

In a similar vein, William Crooke, a young civil servant who was charged with expanding the British colonial rulers' (commonly known as the Raj) knowledge (and therefore oversight) of the peasantry, cataloged the material culture of rural villages in what is now Uttar Pradesh in 1879 (Crooke, 1989 [1879]). He paid a great deal of attention to tools used in herding and milking cattle, but the only utensils he lists as used for milk preparation are the pail and the churn. Among the recorded foods consumed were buttermilk (drunk fresh or hot, or mixed with rice), and maize (a grain native to the New World) or rice boiled in milk, along with milk, *ghee,* and several forms of curds. Overall, Crooke gave scant consideration to food preparation, and one wonders what kind of access he and his Indian assistants had to the kitchen, which was closely guarded against the pollution of outsiders. Nonetheless his observations confirm the primary use of milk for making curd and butter.

Geographic Variation in Milk Usage

Milk, curds, and *ghee* continued to be important elements of South Asian food and drink despite the multiple changes in governance (by Hindus, Buddhists, Muslims, or the British) and influences on cuisine. While their presence in the diet is clearly noted, it is not at all clear how much of them were actually consumed, and thus there is no baseline against which to judge contemporary patterns of intake, or compare with other dairying societies. That said, the diversity of milk uses was remarkable. There did seem to be regional differences in the centrality of milk in the diet across the subcontinent, although with the exception of tribal populations who lived by hunting and gathering or shifting cultivation, milk usage appears to have been ubiquitous. Frederick Simoons, who did foundational work on the geographic distribution of dairying in Asia and Africa, provided numerous descriptions of mostly tribal populations in East, Central, and Southwest India who did not keep dairy animals and did not consume milk (Simoons, 1970). Indeed many of them held quite negative views about milk, in some cases considering it sacrilegious to drink milk, while others reported that they "avoid milk as they feel it makes them bilious" (Simoons, 1970, 565). This range of attitudes toward milk is consistent with the genetic evidence indicating higher rates of lactase impersistence among populations in these regions discussed previously (Baadkar et al., 2012; Babu et al., 2010; Gallego Romero et al., 2012). However, Baadker and colleagues found that individuals who varied in their lactase activity genotype (persistence versus impersistence) did not differ in whether or not they identified themselves as "milk drinkers," defined as those consuming at least 200 ml [<1 cup] of milk per week (2012). This is a very low threshold for defining milk consumption, and so this result is not terribly surprising. Given the overall low levels of milk intake in India (see later in the chapter), and the consumption of milk in small quantities or in fermented form, the geographic distribution of lactase persistence is not likely to be a strong determinant of geographic variation in milk drinking.

While there is a northwest-south cline in lactase persistence frequencies, the historical record does not suggest major differences in attitudes toward milk or consumption patterns except among populations on the fringes of South Asian cultural trends (Baadkar et al., 2012). Indeed, the spread of Hinduism and its associated social structures such as caste has been associated with the incorporation of milk usage among tribal groups, as part of a process the Indian anthropologist M. N. Srinivas (1956) called "Sanskritization." Milk drinking also spread where Indians and/or their religious traditions migrated. For the first millennium CE, dairying was established and milk utilized for both food and ritual in Southeast Asia and Indonesia, but the practice died out as Hinduism and Buddhism were replaced by Islam, suggesting that milk's primary use was ritual.

In South India, the archaeological record shows that cattle were kept as early as 2000 BCE, and in the classical period there is also mention of use of cream, curds, buttermilk, butter, and *ghee,* from both cows and buffalo (Achaya, 1994b). As cuisines developed regionally over the vast subcontinent, milk and milk products were featured more prominently in northern and northwestern cuisines than in eastern or southern cuisines, where fish, rice, and, in the south, coconut milk, were key ingredients in the diet. By the twentieth century though, reports consistently note a stronger emphasis on dairy in the north, with greater production and consumption of milk and other dairy products (Khurody, 1974; Wright, 1937).

Colonial Encounters: Dairy under the Raj

The British presence in South Asia was established in the seventeenth century with the creation of the East India Trading Company, which was disbanded as the Crown assumed formal political control in 1858. The "Raj" lasted until Indian Independence in 1947. During this period the two major milk producing and consuming cultures were conjoined on the subcontinent.

The British colonial experience in India played out quite differently than in North America. This was partly because there existed a strong indigenous dairy culture in India, but also because colonialism took a very different form in South Asia, where the colonists did not establish large enduring settlements or decimate existing populations. As a result, colonialism left a much less tangible legacy on Indian dairy traditions than it did in North America, where there was a substantial replication of European dairying traditions, and those still constitute the core of contemporary American dairy culture.

Perhaps due to the allure of more exotic Indian commodities, there is little mention in colonial writings of the presence of dairy in indigenous cuisine or of milk production, despite the fact that this, at least, must have seemed familiar to the British colonists amidst the myriad aspects of Indian life that seemed strange and quite foreign. Indeed cows roaming in the streets of Indian cities and towns would not have surprised them, as they do many contemporary visitors. At the time, cows were kept in London and other cities, some wandering the streets or grazing in public greens. But with a view toward expanding export commodities and expanding a market for British goods, the new rulers likely overlooked dairy as a largely nonexportable, locally produced foodstuff that was, in the main, prepared in ways that were quite familiar. Furthermore, British rule was punctuated by famines and periods of unrest, causing government institutions to focus on boosting production of grains rather than dairy products. Discussions of food in India during the early twentieth century center around famine, deficiencies in production, and overall food insufficiency, a theme that persists into the twenty-first century.

It is not clear how the British appreciated the role of dairy products in the diet of their Indian subjects during the colonial period, but descriptions of the ways in which they were able to maintain their own dairy-based culinary traditions are available. In a delightful—and best-selling—guidebook for British women posted with their husbands in India, *The Complete Indian Housekeeper and Cook,* Flora Steel and Grace Gardiner devote a chapter to maintaining British dairy traditions in the Indian context (Steel and Gardiner, 2010 [1898]). First, households should maintain their own cows for their milk and dairy products, as those purchased from the market were likely to be adulterated or vehicles of disease (at the time, the same would have been true of milk purchased in English cities). There is no mention of buffalo, so it is reasonable to assume that only cows, more familiar to English households, were kept for milk. However, in contrast to agricultural economists, these authors preferred the "commonest Indian cow" to foreign breeds, remarking on their greater productivity.

Steel and Gardiner provide instructions for various dairy products made in a Anglo-Indian household: butter (clarified, pickled, salted), various cheeses (made from milk or cream, described as float whey, hatted kit, sour cog), and Devonshire cream, as well as numerous other dishes that included milk (puddings, custards, soups, etc.). Much discussion is devoted to the effects of the sweltering subcontinental heat on the production of these products, and how it negatively affected "sweet"—that is, fresh, unsoured—milk. Still, they felt able to produce high-quality cheeses "equal to the best Stilton" (Steel and Gardiner, 2010 [1898], 116). Milk was an essential household product, and Steel and Gardner allocate one and one-half pints per person per day *("N.B. Never stint milk")* and one-half pound of butter for the household (excluding that needed for cooking). Should a home not have its own sources, milk, butter, and cream could be purchased from military or jail dairies, but the quality was suspect and there were especially complaints about the way that butter was produced, with lingering amounts of buttermilk that were deemed unacceptable (Burton, 1993).

While such manuals illuminate the practices of the British in India, they are not particularly informative about Indian diets. Steel and Gardiner do have a chapter on native dishes, and note that a local cheese made from boiling curds (*koya* or *khoa*) could serve as an acceptable substitute for cream, but otherwise they were not inclined, and nor was it their purpose, to draw comparisons between Indian and British uses of milk. One obvious distinction, though, was the conversion of milk into cheese. Britain, like other northern European countries, had a long history of producing a variety of aged cheeses, made by separating curds from whey by use of rennet, an acid derived from the stomach lining of infant calves, lambs, or kid goats, depending on the source of milk. Rennet contains enzymes that nursing mammals use to digest milk, and thus makes an ideal ingredient for cheese making. Rennet was not employed in Indian dairy processing, where bacterial fermentation was used for the production of curd, and, to a much more limited extent, fruit acids (e.g., lemon juice) or soured whey or curds were used to separate the liquid from milk solids. Most people have assumed that the ban on animal consumption would have precluded the use of rennet and hence an elaborated aged-cheese tradition in India. As noted earlier, this prohibition does not have a long history, although it may have contributed to the indifference of Indians to adopting British modes of cheese making during or after the colonial period. Other fresh-cheese industries did develop out of Portuguese influence in Bengal. Bengal became famous for its milk-based sweetmeats such as *barfi, gulab jamu, ras malai,* and other preparations from *chhanna* or *koya*, cheeses that are formed from milk that has been boiled for a prolonged period so that much of the liquid has evaporated.

Tea Culture in India

British cuisine was porous, and several hybrid Anglo-Indian dishes made their way into nineteenth- and twentieth-century British cooking (Collingham, 2006), but British influences on Indian cuisine are most evident in the widespread adoption of tea as a ubiquitous drink. Since tea drunk in both India and Britain is almost always served with milk, and it is currently a major way in which fluid milk is consumed in both places, it is useful to consider briefly the history of tea in the context of colonialism in India. While much has been written about tea—and sugar, as its required condiment in Britain and India—as colonial products (Walvin, 1997), and how the colonial endeavor created both a supply and demand for these exotic commodities (Mintz, 1985), the role of milk, as an indigenous product, has been overlooked as a facilitator of their spread in both countries.

Tea production and consumption were deeply intertwined with the colonial endeavors of the Raj. Indeed it could be said that the advent of tea consumption in India is largely attributable to the British goal of controlling a tea supply. The East India Company, a private company established to bring Asian commodities to the British market in the early eighteenth century, found itself unable to maintain a reliable supply of tea ("an article of the greatest national importance" by the late eighteenth century) from China to meet domestic demand. To offset the trade imbalance with China, the British engaged in illegal trading of opium with China to fund purchases of tea. Protestations by the Chinese resulted in two Opium Wars, which ultimately opened up China to trade. Other nations now had direct access to Chinese tea, sharply cutting into British profits from the tea trade. Furthermore, the East India Company's monopoly on Chinese tea had ended, and it was now interested in establishing tea plantations in the politically tractable areas of South Asia, where wild tea plants had been discovered. However, without knowledge of cultivation and production of comestible tea, early efforts met with more failure than success. The history of these attempts as well as the political intrigue surrounding the tea trade with China has been the subject of several books (cf. Macfarlane and Macfarlane, 2004; Mair and Hoh, 2009; Saberi, 2010). Eventually, using varietal plants smuggled out of China and local indigenous strains, and ascertaining the ecological conditions that fostered the growth of tea plants (in mid-elevation areas of Assam, Darjeeling, and the Nilgiri hills of South India), a viable Indian tea industry was established, and its exports grew as those from China dwindled in the late nineteenth and twentieth centuries.

By the time the British had established political and economic control over much of the subcontinent and established a tea industry there in the mid-nineteenth century, tea drinking had become a routine part of middle- and upper-class life in England, and was rapidly becoming entrenched among the working classes as well. This tradition was maintained among Anglo-Indian households, and provided a comforting domestic presence and assurance of the connection to social life of the mother country. It was in its quintessentially British form—with milk and sugar—that tea was consumed by the British at home and in India. But its adoption by millions of Indian citizens in the late twentieth century may be less of a colonial imitation or legacy than a new flavoring (i.e., tea) for an old drink (i.e., milk, including that mixed with sugar).

In the nineteenth and well into the early twentieth century, tea in India was consumed almost exclusively by expatriates and produced to meet the export market. In 1900 India ranked fifteenth among countries reporting tea consumption, with only 0.04 kg/person/year, compared to the United Kingdom, at 2.8 kg/person/year (Grigg, 2002). Its use among Indians was confined to an herbal remedy—it was too expensive and required utensils that were unavailable to most households: teapots, cups, saucers, sugar bowls, and milk jugs (Collingham, 2006). Coffee had a longer history there, having been introduced by Arabs and Persians in the seventeenth century, but it was not widely consumed. As Lizzie Collingham writes, at the turn of the twentieth century, most Indians did not know how to make a cup of tea and were reluctant to drink one; they preferred water, or in the north, buttermilk (2006). She argues that the conversion of the population to tea drinking was the result of what must have been the first major marketing campaign in India, as the British-owned Indian Tea Association set about extending their overall sales by creating a domestic market for tea. To do so they had to establish a new tea-drinking habit among the Indian population at the turn of the twentieth century.

Purposeful establishment of new food habits—especially in a large, diverse, dispersed, and mostly poor populace with a well-established cuisine—is notoriously tricky, and initial attempts to encourage stores to stock tea and consumers to buy it were disappointing, leading one marketer to comment, "increasing the consumption of tea in India is undoubtedly the most difficult branch of the work" (Collingham, 2006, 195). As is often the case though, exigency provided an opportunity. Factory workers in World War I were offered tea during breaks, and industry owners realized the value in keeping their workers energized. The railway system, a sprawling network of track constructed throughout the subcontinent, provided another venue for tea marketing. The Indian Tea Association sent vendors into major rail stations with kettles of hot tea, hawking *chai* to travelers, and competing with those peddling *pani* (water). Subsequently, the Indian Tea Association organized tea vendors throughout large cities and ports, and the tea shop became a ubiquitous presence in any marketplace, while instructors went door to door demonstrating the art of tea preparation and the joys of tea consumption to housewives.

According to Collingham, "Although the European instructors took great care to guide the tea vendors in the correct way of making a cup of tea, they often ignored this advice and made tea their own way, with plenty of milk and lots of sugar. This milky, intensely sweet mixture appealed to north Indians who like buttermilk and yogurt drinks *(lassi)*. It was affordable and went well with the chapattis, spicy dry potatoes, and biscuits sold by other station vendors, running alongside the carriage windows as the trains pulled into the station" (2006, 196). In other words, Indian *chai wallahs* made tea their own, although much to the marketers' dismay, this mode of preparation required fewer tea leaves. The form in which tea was drunk is reminiscent, in a perhaps more quotidian sense, of the ancient *soma*, which was also a stimulant mixed with milk. Populations of the subcontinent had been drinking boiled milk mixed with spices and sugar for millennia; tea represented one more substance that could be mixed with these, indeed another that also served to keep workers, housewives, and students alert in a rapidly modernizing society: another chapter in the milk-drinking epic.

From this perspective, the Indian *chai* tradition is less an imitation or importation of a British tradition (remembering of course, that this too has a relatively short history in British gastronomic life) than an embracing of a new option for the consumption of this traditional beverage. While the proportions of tea, water, milk, and sugar vary by individual tea preference or economic circumstances, tea may have begun as a flavoring for milk, rather than being something to which milk was added, which was how tea drinking unfolded in the United Kingdom. In India, tea was prepared by boiling pressed tea leaves with water, milk, and sugar instead of being steeped in hot water with milk and sugar added separately to individual taste. Such practices required tea equipage, which was beyond the reach of most Indian consumers.

While the institutionalization of tea drinking occurred during the Raj, only after independence did tea drinking became noticeably more widespread. Until then it was largely an urban middle-class practice, and the vast majority of the tea crop continued to be exported as a substantially higher price could be gotten from British distributors than from local vendors (Lutgendorf, 2012). As one individual from the early 1930s noted, "The first time I had a cup of tea was when I came to Bombay. In the village we used to drink only milk, and water," (Collingham, 2006, 198). Even after independence, tea drinking was not common. Efforts to promote tea domestically began as global tea prices plummeted during the Great Depression. Initially a drink viewed with suspicion (in part because of its link to British imperialism and exploitation), tea marketing messages coalesced around nationalist themes, positioning tea as an indigenous product, one that could "energize" and "awaken" the Indian citizenry to the task of national building (Lutgendorf, 2012). A survey from the late 1970s shows that 50 percent of rural and 66 percent of urban households reported drinking tea at home; somewhat fewer reported consumption outside of the home. A nontrivial portion of income went to tea purchases; around 3 percent of rural and urban incomes was spent on tea for home consumption, while about twice that was spent on tea purchases away from home (Vijayagopalan, 1988). And while at the turn of the twentieth century over 90 percent of tea was exported, by the 1990s only 20 percent was sent overseas, and the domestic market provided the main outlet for the tea supply (Grigg, 2002). Thus over the twentieth century tea became an ingrained beverage in Indian culture, and provided a new way to use milk. I will consider the more recent surge in tea and coffee consumption along with the recent rise in milk intake later in this chapter.

Dairy Trends in the Early Twentieth Century: Laments about Low Production and Consumption

Despite the British predilection for dairy products, they paid little attention to the indigenous dairy economy until the early twentieth century. While in England and in other European countries and the United States dairy industries were rapidly expanding into large-scale commercial operations with widening distribution networks, in India dairying remained a largely domestic enterprise for Indians and the British. By the end of the nineteenth century there were some large herds attached to military establishments to provide milk for soldiers and patients in military hospitals, and some at the few agricultural and veterinary colleges, dairy institutes, and breeding farms. There were also dairies associated with royal residences and a few private firms such as Polson and Keventer that supplied Bombay and Calcutta, respectively, with their milk. Early on, a dairy cooperative model, based on those developed in Germany, was established to more efficiently pool milk from dispersed, small-scale dairy farmers to serve the large urban market. As was similarly happening in the United States and Europe, dairies were also established within growing urban areas to facilitate movement of this perishable product into urban markets. But by the end of the Raj, it is estimated that only 2 percent of India's dairy production came from some kind of organized firm (Achaya, 1994a).

In 1916 the Government of India organized the first census of cows and buffalo, which then took place about every five years starting in 1920.

A flurry of interest in commercial dairy production and processing followed. Pasteurization of commercial milk became standard practice after World War I, and while refrigerated railway trucks became available in the 1930s, it was prohibitively expensive to transport milk this way, so milk remained a largely local, or at most, regional commodity. Cream separators were in use in cities by the early twentieth century for making butter, which also allowed skimmed milk to be sold separately or returned to the producers. Model farms were set up in cities by the Royal Commission of Agriculture in 1928, which was followed in 1929 by the establishment of the Imperial Council of Agricultural Research in Delhi, and then the Agricultural Marketing Department in 1936.

Laments about the low productivity of cows in particular were widespread (Khurody, 1974; Wright, 1937). Even then, production figures largely came from military or government farms where conditions were relatively good and the animals were prime specimens, often foreign breeds or hybrids, but they accounted for less than 1 percent of dairy animals in India. In a 1936 survey of 370 villages in milk-producing regions, cows were found to produce 1.7 kg/day, while buffalo produced twice as much (Wright, 1937). Anglo-Indians commented that an average Bengal cow would yield only one pint of milk per day, necessitating large herds of low-yielding cows, as well as somewhat more productive buffalo (Burton, 1993). Just before Independence it was estimated that there were about twice as many cows as water buffalo, and twice as many goats as cows, but their total production of 180,000 tonnes of milk came from a small fraction of those (Achaya, 1994a). Less than 20 percent of all milk produced reached the market, and the commercial milk supply was comprised of roughly 43 percent cow, 54 percent buffalo, and 3 percent goat milk. Ultimately about half was sold as fluid milk and the other half as milk products, mostly *ghee* (Khurody, 1974), although other authors state that some 70 percent went to *ghee* production (Wright, 1937). Data from 1951 indicated little change in the period right after Independence: cows produced 8,689 million kg (44.3% of all milk); buffalo 10,233 million kg (51.9%); and goats 599 million kg (3.5%) (Sinha, 1961).

The low productivity of dairy animals, was, as noted, a matter of some dismay, and was paralleled by modest intake of milk and milk products. The "Wright Report," a very thorough report on the state of milk production and consumption in India authored by Norman Wright (1937) provides additional insight. Based solely on what would have been available based on per capita *production,* he estimated a per capita intake of 7 ounces [~190 g]/day, inclusive of all dairy products, the lowest of all of the major milk-producing countries (see Table 3.1). While he considered this to be inadequate in terms of the needs of the population, estimates

Table 3.1   Wright's estimates of national milk production and consumption in
             selected countries in the 1930s

Country	Total milk production, 1930–1934 (millions of gallons)	Population (thousands)	Daily per capita production (ounces)	Daily per capita consumption (ounces)
New Zealand	870	1,559	244	56
Denmark	1,200	3,551	148	40
Australia	1,049	6,630	69	45
Canada	1,580	10,377	66	35
Switzerland	607	4,066	65	49
Netherlands	970	7,935	54	35
United States	10,380	122,775	37	35
France	3,150	41,835	33	30
Great Britain	1,474	45,266	14	39
Italy	1,050	41,177	11	10
India	6,400	352,838	8	7

Based on data from Wright, N. C. 1937. *Report on the Development of the Cattle and
Dairy Industries of India.* Delhi: Government of India Press.

from dietary surveys showed substantial variation in levels of intake.
While in the Punjab surveys estimated 14 to 18 ounces [400 to 500g] per
person per day, it was also noted that "the menials use little milk other
than *lassi*. . . . The poorer non-agriculturalists cannot afford to take *ghee*
or milk even occasionally, except on festival days (Wright, 1937, 2).

South India presented a quite different picture, with one survey showing that the majority of families consumed no milk or milk products; the
rest consumed around four ounces [120g] of milk and an equivalent
amount of curds and *ghee* per day. Detailed surveys of milk intake from
households of farmers and laborers in South India were reported in a series
of publications in the *Indian Journal of Medical Research* in the late
1930s (Aykroyd and Krishnan, 1936; Aykroyd and Rajagopal, 1936a,
1936b). While most families owned cows and buffalos and milked
them, the output was very limited and many families sold most of whatever
milk they produced. As a result, over 70 percent of the families consumed
no dairy products at all. The most prosperous households consumed
~0.5 to 14 ounces of whole milk per day, about one-third as much buttermilk, a few ounces of curd, and very few households consumed any *ghee*.

Urban working-class households fared even worse, with an average of
one ounce [28g] of all milk and milk products consumed per day in
Bombay and among jute mill workers in Bengal. Thus collectively, family

budget studies, diet surveys conducted by the Imperial [subsequently Indian] Council of Medical Research, and national sample surveys from the 1930s consistently suggest very low levels of milk intake, ranging between 25 and 50 grams of milk and milk products, including *ghee,* per day (Sinha, 1961).

Consumption seems to have increased somewhat in the 1940s, as estimates from 1945 show an average daily dairy intake of 154 grams, with two to three times that amount in northern states, and much lower levels in the East (Assam's average was 31 g), Central, and South (Achaya, 1994a). In addition to geographic variation, there were also marked differences between rural and urban populations, especially middle-class urbanites. Among households in Lahore (a city now in Pakistan) in 1930, there was a tenfold difference in milk intake between the poorest and wealthier groups, and wealthier households spent around 10 percent of their income on milk, whereas the poorer households spent only around 3 percent. From the late 1930s to 1945, estimates ranged from 150 to 300 g per day for those living in the major cities (Sinha, 1961), compared to the "typical Indian diet" in the 1950s, which contained some 94 g of milk and milk products per day, evenly spread between fluid milk and other milk products (Sinha, 1961). All in all there was an upward trend in milk and milk product consumption as Independence approached, but intake was never particularly high using western twentieth-century standards, with middle-class urbanites consuming the most, but only around one cup per day or its equivalent in dairy products.

One notable trend over during the early Independence period in India was in the way in which milk was consumed. Fresh milk consumption, as a percentage of total milk yield, went up from 28 to 36 percent from 1941 to 1950; *ghee* declined from 57 percent to 43 percent and curd went from 5 percent to 9 percent. In part this is related to the fact that by-products of milk production (e.g., buttermilk) are really rural products and not accessible to urbanites, but it was also due to increased tea and coffee consumption and availability of less expensive alternatives to *ghee* such as hydrogenated vegetable oil *(vanaspati)* (Sinha, 1961). The trend toward greater fluid milk consumption has continued into the twenty-first century.

Regardless of the variation in actual amounts reported above, it is clear that overall intake in India was subject to economic and geographic variation, and, compared to northern European countries, or the United States, Canada, Australia, or New Zealand (all former British colonies, now mostly populated by peoples of European descent) during the same period, consumption was relatively low. It is worth questioning what the norm for milk intake should be, given that most of the world's popula-

tions historically did not consume any milk. In the United States, recommendations during the first half of the twentieth century were on the order of one liter of milk per person per day. In fact, this is the amount that
was available for daily consumption for each individual in the United
States during the 1930s. India's per capita intake seems miniscule by comparison, but the country's dairy production and consumption have long
been compared to the United States or northern Europe rather than other
large countries such as China, where milk had virtually no presence in the
diet until the early twenty-first century (cf. *The Economist,* 1936). Given
the existence of a clear preference for milk and milk products and an
abundance of dairy animals in India, the comparison with the West seems
apt. But it begged the question of why: was it poverty, geography, animal
husbandry, or cultural biases that seemed to limit milk intake?

Wright made India's low consumption levels relative to those of wealthier countries with strong dairy traditions the foundation of his recommendation for increasing milk production as an agricultural priority in India.
He noted that even with a lower "Indian Standard" of 15 ounces [430 g]
of milk (to meet a 16 g requirement for "first-class [e.g., animal] protein"
per capita, which itself was far less than the European standard of 37 g).
The importance of milk to this recommendation was emphasized in relation to the high prevalence of vegetarianism (around one-third of the
population) and overall low meat intake in India, in contrast to Europe
where animal protein was more commonly available and consumed. "In
fact, in a country with as large a vegetarian population as that of India,
the consumption of milk per head should be higher than in countries
with large meat-eating populations, like England and Sweden (*The Economist,* 1936, 627).

While the lower requirement for Indians was admittedly "somewhat
arbitrarily fixed at a low figure in order to make it a feasible standard
under Indian conditions" (Wright, 1937, 4), it was also obvious to Wright
that consumption was tightly related to income. "There is no doubt that
the Indian appreciates the value of milk, and that if he has sufficient means
he will increase his consumption of this food" (1937, 6).

Thus Wright's view, which was consistent with that of many others at
the time, was that the production of milk should be doubled from the current half pound per capita, but he emphasized that this would be ineffec-
tive at boosting consumption unless the price was lowered or incomes
were raised. Similarly, D. N. Khurody, who had long been involved in dairy
marketing in India, remarked in his retrospective book on the Indian
dairy industry that "Our cows and she-buffaloes produce only about
175 kg and 500 kg of milk per annum per animal respectively against

some 3,000 kg in some of the dairying countries abroad. Yet India ranks as the fifth largest producer of milk in the world because of the very large number of cows and buffaloes . . . but they produce less than 7% of world milk production, resulting in a pitiably meager availability of 100 to 102 grams of milk per head per day," (1974, 10). Furthermore, "If one goes through the volumes of literature on dairying and animal husbandry in India, the theme on every occasion has been the same, namely that 'India has too many poor cattle, they produce too little milk, its price for the consumer is very high for the little quantity of the adulterated quality per head that he can get, etc.'" (1974, 7). Thus the poor productivity of South Asian ruminants and poverty came to be seen as the key limitations to what economists and nutritionists viewed as the desirable end of increasing dairy product availability.

Echoing these laments was a 1936 article in *The Economist,* pondering what came to be known as "India's food problem"—insufficient domestic production to feed the large and burgeoning population. The author noted, "Clearly, the most important task confronting the social reformer who seeks to make India's food supply satisfy decent standards of nutrition is to increase milk production in India. The increase has to be as large as 100 per cent. . . . This is not as hopeless a task as may appear at first sight. India has a large stock of milch cattle" (627–628). The author recommended diverting cereals from humans to bovines in order to boost production, a strategy that has remained controversial under more recent milk-production schemes, and among contemporary food systems analysts who see the inefficiencies and higher cost of converting grain to animal products compared to having humans consume grains (or proteinaceous legumes such as pulses or soy) directly (Godfray et al., 2010). There were some recommendations for a milk rationing system: "infants, the old, and the sick ought to get a ration of milk" (Sinha, 1961, 145), but milk did not become part of any Independence-era entitlement programs (see Chapter 5 for the current status of food entitlements in India and milk's absence from these).

Increasing milk productivity remained a government priority after Independence, although greater domestic consumption was not always the explicit justification for expanding the supply. That is, rhetoric more frequently referenced poor production as a problem in and of itself rather than framing the "problem" as too low milk intake. As a result there was no plan to boost milk consumption among those with the lowest levels of intake—the poor. Greater consumption would flow naturally from economic development projects aimed at increasing household income as latent desires could be realized. Thus improving milk production and consumption were bound up in discussions of poverty and projects designed to raise incomes, echoing the ways in which discussions about the poor quality of the urban milk supply were framed in the nineteenth-century United States. These all came together in a dairy development scheme that dramatically altered the dairy landscape in India in the 1970s: Operation Flood.

Operation Flood

In the late 1960s, the Indian dairy industry was targeted for expansion using donations from European countries and the United States under the auspices of the World Food Program (WFP). Europe's dairy industry had been generating massive gains in milk production, which had resulted in a vast surplus at home that threatened to destabilize local milk prices. The WFP was a way of disposing of this surfeit without causing the collapse of local dairy industries in Europe and the United States, and it could do so with an ostensibly humanitarian purpose. Although initially the WFP milk was intended to be used as food aid (mostly skimmed dried milk and butter oil) to bolster the supply and reduce the price of milk in India, the chairman of the Indian National Dairy Development Board (NDDB), Dr. Verghese Kurien, who also happened to be a director of the largest milk producers union in Gujarat, persuaded the WFP to allow his union to reconstitute and sell the milk, with the profits reinvested in expansion of the dairy industry. Thus the European Economic Community found a way to dispose of its surplus milk, and Kurien's group profited from the sale of value-added milk products. The dairy industries of Europe and India, whose markets would have otherwise been inundated with milk, driving down the price of locally produced milk and reducing profits to dairy farmers and processors, both were protected by this arrangement.

This initiated the first phase of "Operation Flood" (OF), a massive dairy development project that has a lasting legacy. Also known as the "White Revolution" as a counterpoint to the "Green Revolution" of the 1960s, which focused on improving grain production, its stated goals were to: make more milk available, at reasonable prices, to urban consumers, including vulnerable groups (namely preschool children and expectant and nursing mothers); enable dairy organizations to identify and satisfy the needs of consumers and producers such that the former's preferences could be fulfilled at a fair price and the latter could earn a large share of the consumer price of milk; improve productivity in dairy farming in rural areas to achieve self-sufficiency in milk, with special emphasis on improvement of small farmers and landless household's incomes; remove dairy cattle from cities; and accelerate development of the dairy economy during and well beyond the program's end (George, 1985). Though originally intended to be a five-year project, it was extended another six years, and then phase II continued for another six years, with an expanded number of cities, a World Bank loan, and the removal of the WFP as the intermediary between India and European countries. Phase III (1985–1996), which also was supported by a World Bank loan, further expanded milk procurement and marketing efforts, and the current six-year National Dairy Plan "Mission Milk" is also financed in large part by the World Bank and aims to increase productivity by developing more productive strains of cows using foreign bull semen.

Kurien became known as "The Milkman of India" and founded the Anand Milk Union, better known as Amul, the most widely known and distributed dairy brand in India with its iconic polka-dotted "Amul Moppet" (the Utterly Butterly Girl), featured in India's longest-running advertising campaign. He also persuaded UNICEF to fund a plant that could make dried milk powder from buffalo milk—existing methods had been developed for cow milk and were not suitable for the higher-fat buffalo milk. This resolved one of the problems related to milk's perishability and opened up new possibilities for production and distribution of buffalo milk within India. Kurien was awarded the World Food Prize in 1989 (funded by General Foods at the time; now supported by Monsanto) for OF, both as a rural development project as well as for its role in making India the largest milk producer in the world, along with several other international awards for these activities. He died in September 2012, and the tributes written in the wake of his death were uniformly laudatory, highlighting his successful establishment of the cooperative model of agricultural production, and its contributions to reducing hunger and poverty in India (although he apparently did not drink milk!).

During his lifetime Kurien was a controversial figure—notoriously blunt and outspoken with a "my way or the highway" attitude—and OF has been the target of numerous critics as well as staunch defenders (Alvares, 1985; George, 1985; Kurien, 1997). My main concern here is its impact on milk consumption in India. Note that increasing consumption—among those with the lowest levels of consumption (the poor)—was only partially a goal of OF: there was no concerted effort to lower the price of milk to make it more widely affordable, and urban populations were privileged, along with a nod to "vulnerable groups." Still, the evidence (including that provided by the NDDB) indicates that OF had little if any impact on per capita consumption, and there is no evidence that it increased milk intake among the most vulnerable (George, 1985). Instead, per capita consumption declined during the early OF years in Bombay, Delhi, and Calcutta, with widening disparities in milk intake between the rich and the poor. The wealthiest 20 percent reportedly consumed 40 percent of available milk, while the poorest 30 percent of the population consumed only 13 percent. In a 2007 World Bank publication Kurien reported a doubling of per capita availability from 112 g to 231 g from 1968–1969 to 2003–2004 (Kurien, 2007), but how this translates into actual consumption across India is not clear. Current estimates suggest that per capita milk available for consumption has increased to 39 liters per year, a 240 percent increase from 1970, but which works out to only ~107 g/day (U.S. Department of Agriculture, Economic Research Service, 2013).

Furthermore, there was no programmatic priority of increasing consumption among villagers, the assumption being that efforts to bolster production would generate more milk for consumption within rural households with milch animals, or more cash from the sale of milk, which could be turned into milk purchases. However, the pricing of milk has remained a delicate issue, with low prices for procurement remaining to encourage milk purchases among urban consumers. Thus rural farmers' incomes did not necessarily rise from participating in the cooperative scheme of OF. Even among villages in the most productive milksheds in Gujarat, consumption remained low; in one such village average consumption was 108 g/day, ranging from 53 g among the poorest families to 299 g among the wealthiest. Moreover, households were encouraged to sell fluid milk, rather than just butterfat for *ghee;* with the sale of butterfat, buttermilk had remained for their consumption, but by emphasizing fluid milk for the market, households lost access to all dairy products, which now had to be purchased from the market.

Current Trends in Indian Dairy Consumption: Modern Milk

The net effect of OF seems to have been to increase milk consumption among the rising urban middle class and wealthy, with little effort to provide it for either the urban or rural poor. As the Food and Agriculture Organization noted in the early 1980s, "For some time to come, milk and milk products will probably continue to be consumed by the more well-to-do people" (Food and Agricultural Organization, 1981, 62). In the 1990s India reversed its post-Independence economic protectionist policies, and opened the door to multinational investment. Several global dairy corporations entered the dairy market: Group Danone, Parmalat, Fronterra, Nestle, and Kraft were among them, although their products are also largely accessible only to urban middle- and upper-class consumers. Along with these new dairy brands are brands of many other beverages (e.g., Coca-Cola, Pepsi, Starbucks) as well as fast food enterprises that compete with vendors of more traditional foods and drinks.

Recent surveys indicate that milk consumption has been increasing in concert with a rise in incomes and purchasing power, particularly among a growing urban middle-class populace (Ali, 2007), just as this became the market for milk in the United States as milk drinking became normative. National data indicate that milk consumption is about 20 percent greater in urban than rural areas, and positively associated with socioeconomic status (Shetty, 2002; Vijayapushpam et al., 2003), a pattern similar to that of the pre-Independence period. The urban middle class has been spending ~10 percent of their total expenditures on milk and milk products, which is more than is spent collectively on cereals and pulses, the mainstays of traditional Indian diets (Saxena, 1996), an amount that is not much different from that in the 1930s. Overall urban and rural expenditures on milk and milk products as a percentage of total food purchases were 16 percent and 13 percent, respectively, in 2010, which represent relatively modest increases since 1970 (National Dairy Development Board, 2012). Milk and milk products are generally considered to have greater income elasticity than other foods in developing countries, meaning that when incomes rise, a greater percentage of them is spent on these products. At the same time, the price of milk has also been going up by about 9.25 percent per year, equivalent to 1.5 percent when adjusted for overall inflation.

Since 1950 the types of dairy products consumed have changed, with increasing expenditures on fluid milk—in the mid-1990s about half of all milk and milk product expenditures were on fluid milk and, as noted earlier, rural procurements were for fresh milk rather than just butterfat. Purchases of milk powder and cheese have also increased, while *ghee* usage has declined (Saxena, 1996). While cheese intake has gone up, it still only represents less than 5 percent of dairy purchases, and so dairy consumption in India remains quite different from that of the United States, where cheese consumption has grown dramatically while milk intake has been declining (Figure 2.3).

At the same time as there were efforts to bolster the dairy industry in India, tea consumption was on the rise, and there are some interesting parallels between the tea and dairy industries. India is the world's largest producer and overall consumer of both commodities, but on a per capita basis the consumption of each is only a fraction of the levels typical

of the United States or Great Britain (Vijayagopalan, 1988). As Vijayagopalan noted,

> Not only has the domestic consumption increased almost three-fold during
> the last decades (1966–1986) . . . but, also, there has been an enormous in-
> crease in the number of households (both rural and urban) who have taken
> to drinking tea . . . although the percentage share of the expenditure on tea
> is considerable for a poor country like India, the apparent per capita con-
> sumption of tea is still low at 0.63 kg per annum. Such a low level of con-
> sumption is all the more paradoxical in view of the fact that India has al-
> ways been the single largest producer and exporter of tea. (1988, 4)

Neither commodity is imported in appreciable quantities (early OF donations aside), but tea, unlike milk, is a major export. However, the domestic
market is the primary destination for Indian tea, with over 70 percent consumed locally (Achaya, 1994b). But, critically, the rise in tea consumption
should be accompanied by a rise in fresh milk intake, unless tea is replacing milk as a drink. Milk is routinely added to tea, and thus is a form in
which many people get milk in their diet, although we don't have very
good estimates of what proportion of milk intake is used in tea. In addition to tea, coffee has become a preferred social drink for young urbani-
stas (Starbucks entered the Indian market in 2012), and is usually drunk
sweetened and with milk.

Considering the trends in Figure 1.7, per capita consumption of fluid
milk surged around 1980–1985, rising to 105 g/day, then dipped down to
about 80 g/day, only to recover between 2005–2010. Current per capita
intake is around 114 g/day, which is roughly a half-cup, although availability was registered at 281 g per capita in 2010 by the National Dairy
Development Board (2012). Of course these are only averages and com-
puted from national level availability and usage data, but nonetheless these
figures indicate a recent rise in milk intake. Survey data on total dairy intake show the same trend (Figure 3.4), and confirm that while average
milk consumption has not increased since the mid-1990s, the percentage
of both rural and urban households that report consuming milk or milk
products has risen (71% of rural households and 85% of urban households,
Ramachandran, 2007).

It is also clear from Figure 3.4 that urban consumption has continued
to outpace rural consumption, although this is not the case among impoverished urban dwellers. Other estimates have been higher, noting per
capita consumption of 121 g milk/day in rural areas compared to 400 g/
day in urban areas at the turn of the twenty-first century (Sharma et al.,
2002). There are also wide regional and state disparities in per capita

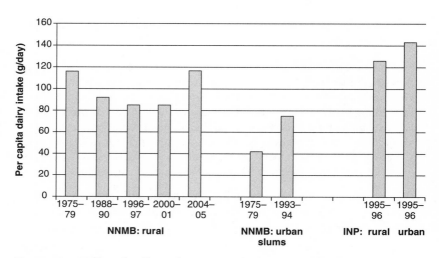

Figure 3.4 Milk and milk product intake, from household consumption surveys, urban and rural India. Based on data from the Nutrition Foundation of India, http://wcd.nic.in/research/nti1947/7.2%20dietary%20intakes%20pr%204.2.pdf.

availability of milk, with the highest levels in northern states such as Punjab (937 g/day), Haryana (679 g/day), and Rajasthan (538 g/day), closely followed by Himachal Pradesh and Gujarat, whereas the eastern states have low availability of milk (113 g in Orissa to 137 g in West Bengal, with much lower estimates among the smaller northeast states) in 2010 (National Dairy Development Board, 2012). Levels in the southern states are between 200 and 350 grams/day. Availability estimates are notably higher than reported consumption levels. In all cases, expenditures on milk and milk products rise with income, and are virtually nil among the poorest households, especially in rural areas (Ramachandran, 2007).

Regardless of the absolute level of intake, all dairy analysts agree that: (1) milk intake has been rising, though it is still well below U.S. or European levels; (2) fluid milk makes up an increasing proportion of overall dairy intake; (3) the demand for milk is rising; (4) the latter is being driven by increasing income and urbanization, and is part of an overall increase in demand for animal products (the so-called livestock revolution evident in many developing countries, Delgado, 2003); and (5) as incomes rise among the poor, their demand for milk will increase and drive consumption trends in the future. In Chapter 5 I discuss how particular meanings of milk in the twenty-first century might also be playing an important role in shaping this increase in consumption in India, but suffice it to say that milk's status as a "traditional" food in India seems to be finding new salience in a rapidly urbanizing population with growing disposable income and expectations.

In sum, India has a deep history of dairy production, and an elaborated set of culinary uses for milk, most of which rely on fermentation. Key themes around Indian dairy traditions have focused on boosting production for India's large population, especially the large numbers of impoverished citizens who are unable to afford any but the smallest quantities of milk. As such, milk is not differentiated from cries to boost production of staple crops for the same reasons, but it does attract special attention as an important animal source food in a country with relatively high rates of vegetarianism (whether voluntary or due to economic limitations) and as an opportunity for rural development, particularly for women. However, there has been little political effort to make milk affordable for the poor or to include it in food-subsidy programs as a "basic" foodstuff, which is one of the starkest contrasts with the United States, where such programs exist in part as a means to absorb the surfeit of milk produced.

Unlike in the United States, where such traditions were entirely derived from European colonization, India's dairy culture is indigenous. Contemporary patterns of tea consumption, which are similar to those of the British, with the inclusion of milk and sugar, represent less of a colonial adoption (although tea is very much a colonial product) than extension of the practice of drinking hot spiced and sweetened milk beverages. As will be detailed in the next chapter, cow milk became a nationalist rallying point in the pre-Independence era, framed by cow protectionist movements, while water buffalo remain the species with greater milk productivity. Milk also has become a means to building larger and stronger post-Independence citizens (Chapter 5).

4

Diversity in Dairy
Cows, Buffalo, and Nonmammalian Milks

A mong countries with large-scale milk production, India is unique insofar as most milk comes not from cows, but water buffalo. Consistent with mid-twentieth-century patterns, buffalo currently contribute 58 percent of the total milk supply (U.S. Department of Agriculture, Economic Research Service, 2013), and the economic literature on India's dairy industry extols the value of buffalo milk compared to that from the indigenous zebu cow. Yet an alternative discourse champions the cow as the superior animal, referencing the sacredness of this bovine to Hindus and its milk-giving properties as an important aspect of its holiness. Within anthropology, the cow's sacred status became the focus of lively and sometimes acrimonious debate that highlighted fundamental theoretical differences within the field. The debate centered on whether the cow's veneration stemmed from religious sentiment, or whether it was—at its foundations—a result of the greater economic value of living—as opposed to butchered—cows. Meanwhile, in the context of this debate, the buffalo somewhat fell by the wayside. With no positive religious significance or strongly held beliefs about it, its value was left to the economists, and the practice of milk drinking was not central to this controversy.

In this chapter I consider the relevance of the debate over the sacred cow to milk consumption practices. Is the cow's sacred status an important contributor to milk consumption, or are decisions about whether to drink milk or what kind of milk to drink, unrelated to this issue? On the U.S. side, no one would argue that the cow achieves the same type of religious significance it has in India, but there is something "sacred cow-ish" about the milk they produce. The term "sacred cow" has become a metaphor for a principle or practice that is inviolable and immutable despite all manner of cultural, political, or economic changes. That is, to challenge milk's central role in the diet, or to acknowledge that milk might not be the "best" food for everyone, is to be at odds with government dietary guidelines, widely employed nutrition education practices, and of course the powerful National Dairy Council, whose fundamental purpose is to increase milk consumption among the American populace. But as we've seen, cows also have cultural significance here, as sources of this idealized food and as icons of an equally idealized bucolic lifestyle with fresh, pure air and food.

The meanings of different types of milk in India and the United States are important influences on the choices people make about what kind of milk they should consume. In India, cow and buffalo milk constitute the two main varieties, with goat milk a distant third. This is partially related to the "sacred cow complex" but also milk's position in Ayurveda, a traditional Indian medical system that has roots in the humoral medical traditions that dominate in Asia and were of historical importance in Europe, as well as to contemporary biomedicine. In the United States there really is no choice when it comes to dairy milks, but there is increasing choice among nonmammalian plant "milks." In both places, health concerns related to milk that influence consumers' choice of which milk to consume focus on its "digestibility." In Ayurveda, for example, cow milk is considered more dilute and hence more easily digestible. On the other hand, both buffalo and cow milk contain the unique milk sugar lactose, and there is variability in the ability to digest this sugar among adults. Plant-based milks contain no lactose, and are marketed at least in part as a milk alternative to those with lactose intolerance. For the first time in U.S. history, viable alternative "milks" are widely available in the marketplace.

Debate over the Sacred Cow

In addition to sorting the world's cultures into the lactophobes and lactophiles, Marvin Harris advocated a theoretical perspective called cultural materialism, which he viewed as a way to understand the "real" causes of what would on the surface appear to be irrational, incomprehensible, or even "silly" beliefs or practices (1979, 1989). In particular, he examined food taboos, which deprive those who observe them from possibly valuable sources of food and nutrients, especially when they are adhered to tenaciously in the face of food scarcity. Two of Harris's most famous examples include the Hindu prohibition of the slaughter and/or consumption of cows and the Muslim and Jewish prohibition of pork. These well-known taboos, stemming from three of the world's largest religions, had been understood primarily as the product of religious doctrines and the cosmologies underpinning them. Although some scholars had proposed that these might serve practical purposes (such as avoidance of pork to reduce the risk of the meat-borne parasite trichinosis), Harris was the first to develop a synthetic theory for human behavior, one that encompassed both the "rational" as well as the seemingly "irrational," and offered a view of human behavior as fundamentally practical.

Cultural materialism holds that societies can be understood as having three components: an economy supporting subsistence, a social organization, and beliefs and ideologies. What Harris proposed was that these components were related to each other, with strong causal arrows going in one direction: from economic to structural to ideological (Harris, 1979). That is, social organization and ideologies stemmed from economic constraints and necessities rather than those driving choices about subsistence. Ideology could certainly influence economic activities, but its causal force was much weaker than that exerted by the economy on ideology. Overall then Harris predicted that while such taboos and their associated belief systems might appear to be bizarre, in fact it is expected that human behaviors ultimately enhance, rather than diminish, human well-being.

In his seminal article on why the cow is sacred to, and its consumption eschewed by Hindus, Harris began with this approach: "I believe the irrational, non-economic, and exotic aspects of the Indian cattle complex are greatly overemphasized at the expense of the rational, economic, and mundane explanations" (1966, 51). He argued that the ideological principle underlying the cow's sacred status, *ahimsa,* or nonviolence, is insufficient to understand why cows are not slaughtered or eaten. Instead, he argued that cows and humans have symbiotic relationships in the South Asian ecosystem, insofar as cows have great economic value as sources of dung (for fertilizer and fuel), traction (bullocks), beef (for lower castes or non-Hindus), and leather. Milk was noted, but considered a relatively minor contribution by cows, since buffalo were preferred due to their greater productivity. In addition, because they are ruminants, cows do not compete with humans for food (reports of their destruction of crops notwithstanding, Harris, 1966). Because of Westerners' biased view of cattle only as sources of meat and milk, Harris argued that these other benefits have been overlooked, such that the practice of keeping seemingly unproductive cows alive seems irrational.

The commentators on Harris's original article were generally favorably disposed to his argument, but subsequently both his overall approach and application of it to the Indian sacred cow issue were subject to often vitriolic critiques, which centered on (1) discrediting the ecological/economic value and emphasizing the cost of excess cows; (2) the political importance of cow protection; (3) the historical emergence of the sacred cow within Hinduism, its role within the Hindu cosmology, and the religious significance of *ahimsa;* and (4) why cows needed to be given sacred status to be protected from slaughter. With regard to the latter, Harris's argument is that sacred status of the cows removes the temptation to kill and eat them. Why such an enticement would exist is not clear, other than a sense that humans have a predilection for meat and beef in particular (which itself may reflect a Western bias). A similar charge could be made against one of his most vociferous critics, the geographer Frederick Simoons, who lamented the "waste" of beef in a population reportedly experiencing widespread protein deficiency (1979).

Religious Perspectives

Frank Korom has argued that the cow serves as a "key symbol" uniting the disparate beliefs and practices that characterize Hinduism, and that cow worship is the expression of "a central belief that the cow is good, whole, pure, and embodying all aspects of the cosmos within her" (2000, 190). This belief may share common roots with eastern Mediterranean cattle cults documented from ~6500 BCE, where both bull deities and mother cow goddesses were worshipped (Lodrick, 1981). In Hinduism the cow represents the entirety of the universe (Figure 4.1) as it is constituted in space and time. Each cow is an embodiment of infinite deities.

Cow protection is underpinned by the principle of *ahimsa,* and the institution that realizes this ideal is the *goshala* ("a place for cows"), a home for aged or disabled cows. Sponsored by temples, trading castes *(vania),* Gandhian ashrams, Hindu educational centers, or government officials, *goshalas* have diverse forms and purposes, but all recognize the religious significance of cows and provide resources for their care. Many of the cows are beyond their economically useful days, but others are kept precisely for their milk production, either for ritual purposes or for local consumption (Lodrick, 1981). Most *goshalas* take only cows, although other institutions, such as *pinjrapoles,* take animals of all kinds and are dedicated to the principle of *ahimsa* in its broadest sense. Villagers may be able to deposit cows there that they cannot afford to keep, and have them returned when their fortunes improve, suggesting that *goshalas* function to reduce the costs of maintaining animals during difficult times. *Goshalas* are found throughout India, but are concentrated in the

Figure 4.1 Kamadhenu and her calf. The celestial cow has a woman's face and hair, multicoloured wings, the tail of a peacock and two cow's tails. Her belly contains the ocean, here only summarily sketched. Her diminutive calf, an exact replica of its mother, is shown underneath her belly. Tiruchirapalli, ca. 1830.
© Trustees of the British Museum.

Indo-Gangetic Plain, especially in the state of Gujarat. Lodrick (1981) posits that this geographic distribution reflects the influence of the Vedic tradition in the north, which was not part of Dravidian cultures indigenous to the south or tribal cultures in the north and east. It is worth noting that Gujarat is currently the center of milk production in India, and also where many Jains live. For Jains, the cow is just one animal deserving of sanctity, and all animals, from insects to bovines, deserve protection.

Cows are symbolically linked to the earth and mothers, which both have life-giving properties (Korom, 2000). The five products of the cow (milk, curd, *ghee,* urine, and dung) are considered inherently pure and are used to maintain the purity demanded for religious practice and caste separation. Hindu scriptures identify the cow as the mother of all civilization, its milk nurturing its citizens. The cow emerged as a symbol of fecundity, maternity, and life-giving sustenance during the Vedic period, but evidence of such status was notably absent (aside from bulls on seals) in the earlier Indus Valley civilization. In contemporary Hinduism the bull (Nandi) is considered the vehicle of Shiva, the god of fire, fertility, and the Great Destroyer, and large Nandi sculptures are found outside of Shiva temples. Importantly, while the "cow as mother" motif is a key theme in Hinduism, there is no—and apparently never has been—a cow goddess, and sculptural representations of cows are scarce relative to those of bulls. The closest thing to a cow goddess is Kamadhenu. Shown in Figure 4.1, she is the divine cow that is considered the mother of all cows. She is the "cow of plenty," and worshipped indirectly through veneration of cows. Nandini is another name by which Kamadhenu is known, and is the name of a major milk brand in South India.

Female cows are more commonly associated with Krishna, an avatar of Vishnu, the preserver of the world. Krishna is often represented as a mischievous child-god with his hand in the butter jar (Figure 5.2 in the next chapter) or as the Divine cowherd *(Krishna govinda),* consorting with the *gopis* (cow maidens; see Figure 4.2). As an incarnation of Vishnu, Krishna is especially worshipped by Vaishnavites, who are frequently part of the Vaishya varna of the caste system and are traders, merchants, and farmers. But Krishna has pan-India popularity, and according to Lodrick his divine child form "holds a tremendous popular attraction in a society traditionally viewing the prime role of women to be the bearing of male children" (1981, 66). Thus, while cows may be revered and have close associations with male Hindu deities, they are not really the central focus; the association of bulls with Shiva and cows with Krishna may serve to dilute the female, milk-giving properties of cows.

Figure 4.2   Krishna relaxing with the gopas and gopis, Punjab Hills, Pahari School, ca. 1730–1740. © Trustees of the British Museum.

Political Uses of the Cow

Movements to strengthen the protections extended to the cow gained their impetus in the nineteenth century in the writings of Swami Dayanand Saraswati, founder of the Hindu reform organization Arya Samaj. In *Gokarunanidhi*, Saraswati framed cow protection as essential to health and prosperity, invoking the concept of *dharm (dharma)* in the sense of the natural order of things. As Cassie Adcock maintains, "*Dharm* was translated throughout in terms of the common good, measured in material terms and produced through harmony with scientific and natural principles," rather than referring to a "religious" doctrine per se (2010, 303). The argument was made that "to kill cows brings harm to all the world *(samsar)*. When cows are protected, milk is abundant and therefore inexpensive; dairy products are accessible to the poor, who eat less grain; all people produce less excrement, meaning fewer 'insalubrious gases'; the result is a decrease of disease and an increase in the happiness of all" (Adcock, 2010, 302). During the time of the Raj, appealing to religious sentiment could not be the foundation for a petition to protect cows due to the ostensible commitment to the principle of religious toleration, but employing the concept of *dharm* allowed cow protection to be read in a variety of ways, depending on the audience. Given other writings of the Arya Samaj that derided the British, Muslims, and low-caste groups, all of whom could be charged with eating beef, the plea for cow protection could easily be—and often has been—read as an attempt to assert a pan-Hindu identity and political agenda (see Figure 4.3).

Cow worship and protection gained force during the period of British colonialism in South Asia. According to William Gould, in the politics of late colonial India "the concept of 'mother cow' could be twinned with the often used depictions of 'Mother India' and the life-giving, pure qualities of cow's milk could be associated in the minds of audiences with the purity and strength of the nation" (2004, 78). The imperial government, coming from a culture that viewed cows solely as a source of dairy products and meat, showed little sympathy for cow worship, and cow protection became a rallying cry against British rule, much as it had been used by the great Marathi leader Shivaji in his attempt to take back power from the Muslim Moguls during the medieval period. Mahatma Gandhi made cow protection part of his agenda, noting "The central fact of Hinduism is cow protection. . . . Hindus will be judged not by their *tilaks,* not by the correct chanting of mantras, not by their pilgrimages, not by their most punctilious observance of caste rules but by their ability to protect

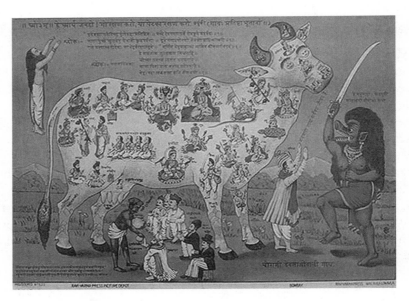

Figure 4.3   "The Cow with 84 Deities" [Kamadhenu], ca. 1912, Ravi
Varma Press. Part of anti-Muslim propaganda by cow protection groups
to protest beef eating during the Raj.

the cow" (Gandhi, 2001, originally published in *Young India*, 6 October
1921). Indeed, he went on to say:

> Mother cow is in many ways better than the mother who gave us birth. Our
> mother gives us milk for a couple of years and then expects us to serve her
> when we grow up. Mother cow expects from us nothing but grass and
> grain. Our mother often falls ill and expects service from us. Mother cow
> rarely falls ill. Hers is an unbroken record of service which does not end
> with her death. Our mother when she dies means expenses of burial or cre-
> mation. Mother cow is as useful dead as when she is alive. We can make use
> of every part of her body her flesh, her bones, her intestine, her horns, and
> her skin. Well, I say this not to disparage the mother who gives us birth, but
> in order to show you the substantial reason for my worshipping the cow.
> (Gandhi, 2001, originally published in *Harijan*, 15 September 1940)

Thus, for Gandhi, cow protection was an essential and fundamental
aspect of Indian life; he emphasized both the economic and spiritual benefits of cows, which provided a justification for their protection. This
view was invoked in an effort to unify and motivate resistance to British
rule, which had allowed cow slaughter, and thus served explicit political
purposes as well. The cow became a potent symbol of "Mother India" in

Figure 4.4   The cow as Mother India: "Her precious milk nourishes Hindus, Muslims, Sikhs, and Christians alike." Image courtesy of Patricia Uberoi.

the drive for independence from England during the late colonial period, and as Figure 4.4 shows, cow milk could also be considered pan-religious, uniting the diverse faiths in India in the effort to present a common value (Uberoi, 2003). Given the waxing and waning of cow protection and injunctions against meat eating over the course of South Asian history, deployed by those in power (or those seeking to establish or extend their power) to exert control, some scholars maintain that cow worship has long had political motivations (Diener et al., 1978).

The question of whether cow protection was a utilitarian or religious issue has been at the forefront of political debates in India. It dogged leaders charged with formulating the Indian Constitution, which was founded on the principle of secularism. Cow protection was embedded in the Indian Constitution after much debate over whether this was appropriate in a secular state. It appears in the Directive Principles to States, which are nonenforceable guidelines for laws and policies for national and state governments. The directive reads: "It [the state] should also organise agriculture and animal husbandry on modern and scientific lines by improving breeds and prohibiting slaughter of cows, calves, other milch and draught cattle." As stated, cows are not singled out for protection (as a religious motivation would have provided), but rather the economic utility of bovines in general was emphasized by the broadness of the ban. Thus the very question of whether the sacredness of the cow is a religious or economic issue was front and center in constitutional deliberations. States have moved forward with their own laws, some of which are more restrictive and encompass all zebu and buffalo, while others have singled out cows but made no mention of buffalo. Cow protection/slaughter continues to be a volatile political issue.

Villagers' Perspectives

The issue of cow worship is not easily resolved with one theoretical perspective. With an animal of such fundamental importance to early South Asian inhabitants and the high esteem in which they were held (Batra, 1986; Korom, 2000), the economic and spiritual attributes of the cow undoubtedly became intertwined and maintained by a mixture of economic, political, cultural, and religious structures and sentiments. That said, whatever beliefs Hindus have about the holiness of cows does not result in a willingness to encourage or even allow their unfettered reproduction. There is clear evidence of the passive, but quite intentional, culling of bovines and careful management of herd composition.

As ethnographers of North India Stanley Freed and Ruth Freed observed, "The proposition that an idea as powerful as the Hindu belief in the holiness of the cow has no effect upon how Hindus treat their cattle seems contrary to common sense" (Freed et al., 1981, 489). Yet they and others have observed that the sex ratio of cows is not what would be expected if all were cared for or even allowed to live. Bullocks (castrated bulls) outnumber females by a wide margin, especially in North India, where bullocks are commonly used in wheat and pulse agriculture (Freed et al., 1981; Vaidyanathan et al., 1982). In contrast, female buffalo are more common than males and used primarily for milk.

In the south and east, where rice is cultivated, male buffalo are more useful for traction in flooded fields and, not surprisingly, male buffalo predominate and there are more female than male cows (Vaidyanathan et al., 1982). Female buffalo are more common in areas and households where there is a sufficient surplus of fodder for high-quality buffalo milk production. Furthermore, if the sacred nature of the cow leads to its maintenance, one would expect that Hindu households or villages would have more cows than those inhabited by Muslims or Christians. In fact, the opposite appears to be true (Vaidyanathan et al., 1982). Morgan MacLachlan concluded from this report and a census of bovines in a Karnatakan village, "People in both places [villages in north and south India] adjust cattle holdings to techno-logical advantage by interpreting their religious responsibilities to cattle in view of the economic responsibilities of cattle to humans" (1982, 377).

In sum, the sacredness of cows, laws protecting them, and religious rituals involving their products, including milk, curd, and *ghee* were not evident in the earliest Vedic literature, but emerged over the course of Indian history and have served different purposes in light of social and political changes. These elements have salience to the everyday lives of Hindus, and religious sentiments about the cow have been continually invoked for political purposes.

Is the cow's sacred status is crucial to understanding milk consumption practices in India? Milk production is just one of many services cows provide, and in villagers' eyes, not usually the most important of them (Freed et al., 1981; Harris, 1966; Vaidyanathan et al., 1982). However, in the political arena, the cow's role as a producer of milk has been trumpeted by some factions. There's no doubt that cows have been appreciated as a source of milk since the Vedic period, and as we've seen, milk was a critical component of the diet among these early pastoralist groups. Certainly in some places such as large *goshalas*, milk production for local consumption has been stressed, and some have become sites of cattle breeding and improvement for milk production (Lodrick, 1981). But the relationship between the sacredness of cows and milk consumption remains opaque. After all, in European and among other pastoralist groups in Africa and Asia, cows have not been ascribed sacred status despite their milk being a fundamental food source. Recognition of economic utility, valuation of dairy products, and sacredness do not go hand in hand across dairying populations; indeed, the first two go hand in hand with slaughter and consumption of animals also used for dairy.

What to Drink...Cow Milk or Buffalo Milk?

The role of milk in the sacred cow debate is complicated in India by the fact that the water buffalo is more highly valued for its milk, and most milk available comes from buffalo, even though there are 26 percent more female cows than female buffalo (IUF Dairy Division, 2011). Although hybrid cows can produce as much milk as a buffalo, indigenous zebu cattle generally have a yield that is half as much as buffalo (India Study Team, 2007). Most commercial milk is an unspecified blend of milk from cows and buffalo. As Table 4.1 shows, buffalo milk has almost twice the fat of cow milk, and milk's cost is primarily determined by its fat content (Wattiaux, 2011). Milk fat can be converted into *ghee* and preserved indefinitely. *Ghee* is considered a pure substance, both in the idioms of microbial resistance and ritual purity, and foods cooked in *ghee* are protected from caste contamination and can be shared across household and caste lines (Mahias, 1988). Thus while cow protection efforts have emphasized cows as sources of valued milk, the fact that buffalo produce the majority of milk and milk fat, and that buffalo are prized more for their milk than are cows has been largely ignored.

This fact was not lost on Gandhi, who lamented his compatriots' preference for buffalo milk.

> I am amazed at our partiality for buffalo milk and *ghee*. Our economics is short-sighted. We look at the immediate gain, but we do not realize that in the last analysis the cow is the more valuable animal. Cow's butter (and *ghee*) has a naturally yellowish colour which indicates its superiority to buffalo butter (and *ghee*) in carotene. It has a flavour all its own. Foreign visitor's cow's milk they get there [*sic*]. Buffalo milk and butter are almost unknown in Europe. It is only in India that one finds a prejudice in favour of buffalo milk and *ghee*. This has spelt all but extinction of the cow, and that is why I say that, unless we put an exclusive emphasis on the cow, she can not be saved. (Gandhi, 2001, originally published in *Harijan*, 8 February 1942)

Gandhi framed the superiority of cow milk in both nutritional terms (although his analysis was incorrect; buffalo milk has more of the caro-

Table 4.1 Nutrient composition of buffalo, cow, and human milk, per 100 g

	Cow	Buffalo	Human
Water (g)	88.0	84.0	87.5
Calories (kcal)	61.0	97.0	70.0
Protein (g)	3.2	3.7	1.0
Fat (g)	3.4	6.9	4.4
Lactose (g)	4.7	5.2	6.9
Minerals, including calcium (mg)	72.0	79.0	20.0

Based on data from Wattiaux, M. A. 2011. Milk composition and nutritional value. In Babcock Institute (ed.), *Dairy Essentials*. Madison, WI: Babcock Institute for International Dairy Research and Development.

tene converted into the fat-soluble vitamin A, which is likely related to the higher fat content of buffalo milk) and as a religious mandate. Curiously, as Gandhi was a champion of indigenous Indian village traditions, he also intimated that European usage of cows was another justification for the consumption of cow milk and the protection of cows. This sentiment is echoed in current cow protection rhetoric, as described in a 2008 article in *The Indian Express*. In Madya Pradesh, a government-sponsored cow donation program butted heads with the state's dairy federation, which insisted that there was a much greater market for buffalo milk than cow milk. Yet the government commission, under the political leadership of the BJP, a conservative Hindu nationalist party, balked. As the report noted, "When the commission learnt that cow's milk would be mixed with buffalo's and then sold, its chairman Babulal Jain put his foot down. "Cow's milk is the elixir of life. Those who consume it become energetic and smart unlike buffalo's milk, which produces lazy people" (Ghatwai, 2008).

There is an active anti-buffalo-milk consumption campaign in India that asserts that cow milk is superior to buffalo milk, especially in terms of its effects on mental abilities. At www.cowpower.info, a site devoted to the Gau Maata (mother cow), it is asserted that since the countries with the highest levels of cow milk production are wealthy, with high standards of living and low levels of corruption, and that those with the highest levels of buffalo milk production have low incomes and standards of living as well as high rates of corruption: "we can thus conclude that consumption of Cow milk by a nation, has a direct impact on the mental abilities, prosperity and social behaviors of a society." However, as seen in Chapter 3, the National Dairy Development Board in India continues to pursue greater milk productivity among both cows *and* buffalo in the new "Mission Milk."

All of this points to a fundamental contradiction between the religious significance of the cow and the economic superiority of buffalo when it comes to milk. This conflict also reveals itself in the better care received by female buffalos than cows. As Hoffpauir notes, "the best cared for animals are those which are contributing the most to the farmer's economic well-being, namely zebu bullocks and milk buffalo. The zebu bullock provides the indispensable work force required for the cultivation of the crops, and the she-buffalo is a prize possession supplying the farm family with high-fat milk, the *ghi* from which provides the family with a supplemental source of income" (1982, 225). This better treatment and economic valuation does not extend to male buffalo (except in the south), nor does it protect the buffalo from slaughter in many states. In fact, buffalo sacrifice continued until recently and is still practiced in some areas (Jha, 2002).

Buffalo themselves are considered inauspicious: "unclean, unlucky and a bad omen . . . associated with death, disease, and demons" (Hoffpauir, 1982, 227). In Indian mythology, evil is often represented by water buffalo, which also serve as the vehicle of Yama, the Hindu god of death (see Figure 4.5). In contrast to the cow, within whom all the gods dwell, Yama's fortress is filled with buffalo (Hoffpauir, 1982). Given these associations, it comes as no surprise that buffalo are not to be found in contemporary milk advertisements, even though generic commercial milk is more likely than not to contain buffalo milk. This is not widely broadcast; instead happy cows predominate in milk commercials, and milk brands use traditional names: "Mother Dairy" is a pan-Indian brand owned by the National Dairy Development Board (and a product of Operation Flood) that conjures up both the sacred cow-as-mother and the nationalist "Mother India" motifs, much like the earlier attempts to link Indian citizens through a common cow-milk-drinking experience (Figure 4.4). Likewise Nandini, the brand of the South Indian Karnataka Milk Cooperative Federation, harkens back to the wish-granting cow Kamadhenu. However, as Figure 4.6 indicates, the cows featured in such ads are not the indigenous beloved zebu cattle that serve as potent religious symbols, but rather foreign breeds such as the distinctive black-and-white Holstein. Thus the zebu cow and milk's historical local meanings in India are

Figure 4.5   The water buffalo is Yama's vehicle. © Trustees of the British Museum.

Figure 4.6   Nandini brand milk advertisement, using a Holstein cow. http://
www.kmfnandini.coop/.

not being harnessed to sell milk: Western cows with their greater size and
productivity are used, suggesting that, at least among the middle class, to
whom these advertisements are aimed, the traditional zebu cow motif is
not sufficiently compelling. Given widespread recognition of the poor
state of zebu cattle and their milk-producing capabilities, such images
would not represent a thriving industry (or thriving consumers) amidst
India's current economic boom (see Chapter 5).

What is also curious is that buffalo milk, milk products, or *ghee* can be
used in Hindu religious rituals—there is apparently no bias against it or
for cow milk usage in this context (Ferro-Luzzi, 1977; Nanda and Nakao,
2003). Indeed when I have asked my Indian colleagues about a preference
for cow milk for religious rituals, they are often puzzled—it is the milk or
milk products that are important, not the source. This is in part because
milk—regardless of the source—is considered already cooked. As such it
is considered pure, and foods cooked with milk, curd, or *ghee* are thereby
purified. "Milk is none other than the sperm of Agni, god of fire, and
anything that comes from Agni is by its very nature cooked . . . and so,
whether the cow be black or red, her milk will be white life fire" (Mahias,
1988, 269). Furthermore, Agni and Yama are portrayed as friends in the
Rigveda. However, Hoffpauir suggests otherwise, based on his work on
buffalo in India in the early 1970s. He argues that buffalo milk has no
inherent religious qualities, unlike zebu milk, which has a long history of
use in ritual offerings and purifications (Hoffpauir, 1977). In contemporary urban contexts, where milk is procured from commercial sources
and its provenance is uncertain, it may be that the distinction between
cow milk and buffalo milk for ritual purposes has eroded. However, the
ubiquity with which my question has been answered suggests a more
generalized apathy toward the source of milk, at least with respect to
this usage, perhaps due to the linkage between Agni and Yama (see below

for clear distinctions made in health contexts). Furthermore, from an etymological perspective, there is no difference, linguistically, between "cow milk" and "buffalo milk." Even the words for "cow" appear to mean a generic "female bovine creature" (Rebecca Manring, personal communication).

Many contemporary commercial dairies do offer a separate cow milk, but without any religious references. Mother Dairy notes only that its cow milk is "easily digestible," which references Ayurvedic understandings of milk (see next section). Nandini likewise has a separate cow milk, marketing it as "the good life," with an image of a Holstein cow. Milk is often advertised as "pure," which serves a double purpose of emphasizing its safety and its religious significance. However, since "pure" is used for all milk products rather than just cow milk, it is more likely simply a guarantee against contamination or adulteration. Nonetheless, the fact that large dairies offer both kinds of milk suggests that there is a market for pure cow milk. It is telling that there does not seem to be a separate marketing strategy for exclusively buffalo milk.

While it might seem surprising that the Indian dairy industry does not make greater use of traditional motifs, the fact that water buffalo milk production predominates due to greater productivity is but one obstacle. India has a sizeable Muslim population (~140 million; ~14% of the total population) and strong nationalist Hindu political movements. The dairy industry must be careful not to incite communal antipathy by reference to religious symbols in its marketing. One solution has been to divorce milk from its traditional roots and highlight nutritional science perspectives that emphasize milk's nutrients, while playing on some familiar or unifying nationalist themes (Wiley, 2007).

Gyorgy Scrinis (2013) coined the term "nutritionist" to refer to the privileging of nutrients as descriptors and markers of food value. As he notes, nutritionism is "characterized by a reductive focus on the nutrient composition of foods as the means to understanding their healthfulness, as well as by a reductive interpretation of the role of these nutrients in bodily health" (Scrinis, 2013, 2). Nutritionist messages have become globally ubiquitous, and in this case, serve to link Indian consumption patterns with those of the wealthy global north, and provide additional justification for milk's centrality in the diet. By and large, the meanings of milk portrayed in milk promotions are squarely in this vein, with myriad references to milk as a rich source of protein and calcium and essential to building strong bones (see Chapter 5). Such messages do not justify a bias toward cow milk except insofar as saturated fat is deemed a "bad" nutrient and buffalo milk has more of this. However, with the ubiquitous availability of tonned (skimmed) milk and the traditional removal of fat to make *ghee* mitigate this potential concern. I will return to this concept in a following section that considers the ways in which milk is understood and promoted in the United States.

Ayurvedic Understandings of Cow Milk and Buffalo Milk

Ayurveda provides an interpretive framework through which consumers understand the potential effects of different kinds of milk on the body. Ayurveda (the "science of longevity") is a pan–South Asian medical system that originated in the Mauryan period (second and third centuries BCE) and developed through the seventh century CE. The foundational texts are, in chronological order: *Charaka Samhita, Sushruta Samhita,* and the *Ashtanga Hridaya and Ashtanga Sangraha.* In Ayurveda, the body is conceived of as having three semi-fluid *dosas* (*doshas*; lit. faults) or humors: *vata* (wind), found mainly in the large intestine and involved in respiration; *pitta* (bile), localized around the navel and involved in digestion; and *kapha* (phlegm), which is localized to the chest and involved in structural integration (Fields, 2001; Wujastyk, 2001). Individuals vary in terms of which *dosa* predominates in their bodies, and the relative power of each *dosa* collectively determines their "heat" or "coolness." Furthermore, the relationship among the *dosas* has a profound influence on an individual's health—*dosas* out of place, or in too large or small quantity, become irritated.

Among those things that can disturb the *dosas* are changes in weather/ season, emotional distress, sins of various kinds and inappropriate behavior, and diet (Wujastyk, 2001). Given that the primary activity of the body is considered to be digestion (conceived of as cooking, or the digestive fire [*agni*]), dietary regimes are of the utmost importance in Ayurveda. Foods have specific effects on the *dosas,* and depending on an individual's constitution (i.e., the predominance of one *dosa*), diets with different quantities and types of food should be followed.

Milk is mentioned extensively in the *Charaka Samhita,* and is used in a wide variety of nondietary therapies, including enemas, herbal decoctions, rubs, bathing, poultices, and so forth. In the diet, milk and its products are considered a "food group," one among grains, legumes, meat, vegetables, fruit, salads, alcoholic drinks, water, sugarcane, condiments, and "dietary preparations." Depending on the type of milk it may be considered light or heavy, although in general milk is light, making it suitable for routine consumption. Cow milk is considered the best of all milks, while sheep's milk is the least desirable. Described as "sweet, cold, soft, unctuous, viscous, smooth, slimy, heavy, dull and clear. It is wholesome, rejuvenating and strength promoting. It promotes intellect, longevity, and virility. Thus cow milk is nourishing, healing, remedial and is used as Rasayana. Warm milk of the cow immediately after milking promotes strength. It is like ambrosia and alleviates all three *doshas* and stimulates digestion. However, cold milk *(dhara sita)* aggravates all three *doshas*. Thus milk should be taken warm" (Guha, 2006). Cow milk promotes the health of all tissues *(dhatus)*. Buffalo milk is heavier, sweeter, and more cooling than cow milk, and because of its high fat content, it impedes digestion and the flow of energy through the body's channels, and is thus recommended for those with "excessive digestive power."

Throughout the *Charaka Samhita* milk is mentioned in many contexts, but its source is not usually specified. Similarly, in the later *Sushruta Samhita,* milk's usages are legion, and only in specific cases is the type of milk noted. Milk of all kinds is described as "the best of all nutritive substances (literally life giving). . . . And since milk is kindred in its nature to the essential principles of life and so congenial to the panzoism of all created animals, its use can be unreservedly recommended for all," (Bhishagratna, 1911, 430). Furthermore, milk in general is "sacred, constructive, tonic, spermatopoeitic, rejuvenating, and aphrodisiac. It expands the intellectual capacities of a man . . . increases the duration of life, and acts as a vitaliser. It is an emetic and purgative remedy, and imparts a rotundity to the figure, and which through its kindred or similar properties, augments the properties of bodily albumen *(Ojas)* and is the most complete and wholesome diet for infants, [and] old men" (Bhishagratna, 1911, 431).

The properties of different mammalian milks are then noted: "The milk of the she-buffalo is sweet in taste, tends to impair digestion and increases the slimy secretion of the organs. It is heavy, soporific, cooling, and contains more fatty matter than cow's milk" (Bhishagratna, 1911, 432). Cow milk "does not set up or increase the normal quantity of slimy secretions in the internal channels of the body. . . . It is cold and sweet both in taste and chemical reaction. It subdues both Vayu [Vata] and Pittum [Pitta] and is accordingly one of the most efficient of vitalising agents" (Bhishagratna, 1911, 431). It might seem that cow milk is the ideal milk in this text, but goat milk actually comes in for even higher praise: "The milk of a she-goat is possessed of qualities similar to those of a cow. [It] proves curative in all diseases owing to the smallness of her limbs and her agile habits, as well as for the fact of her drinking a relatively less quantity of water and living upon bitter and pungent herbs" (Bhishagratna, 1911,

431). However, the section describing the various milks and their preparations concludes with "[cow milk] is the best of all kinds of milks described" (Bhishagratna, 1911, 440). Preparations from milk (curd, whey, butter, *ghee*) have the properties ascribed to their milk source, and curd made from cow milk is singled out as "the best in virtue and quality" (Bhishagratna, 1911, 436).

In sum, the Ayurvedic texts ascribe different qualities to cow milk and buffalo milk, with a bias favoring the former, but in many cases milks are interchangeable and overall are considered among the best and most nourishing of foods. In the Ayurvedic idiom, milk is cooling to the body and a source of vitality, strength, and mental and physical stamina. Milk has profound effects on digestion, with cow milk facilitating digestion (even being used as an enema or purgative) and buffalo milk requiring greater digestive abilities largely due to its higher fat content. Buffalo milk increases *kapha* (phlegm) and induces sleep, while both milks (and curd made from them) subdue *pitta* and *vata*, suggesting soothing effects on digestion.

Gandhi and contemporary advocates of cow milk drew on these Ayurvedic distinctions in their disdain for buffalo milk, but it is worth keeping in mind that the original texts do not indicate such a strong prejudice against buffalo milk. Cow milk is considered "best," but buffalo milk shares many of the same virtues, including salutary effects on intellectual capacity. Furthermore, while the cow produces the five sacred products (milk, curd, *ghee,* urine, dung), and these are also prescribed in the Ayurvedic texts, urine from other dairy animals, including buffalo, is also mentioned. That said, given the ascription of soporific qualities to buffalo milk, it might easily be seen as inducing lethargy, both physical and mental, which in popular view reflects the personality of the buffalo itself.

Given the variation across India in the genetic trait that governs lactase production (and hence milk digestion) in adulthood, and that this likely existed during the time the authors of Ayurveda were working, it is notable that individual variation in the ability to digest milk is not mentioned by them, although such differences would be easily framed in terms of the *dosa*s, with individuals varying in terms of the predominance of *pitta* or *vata*. Buffalo milk is contraindicated for those with *kapha* tendencies, but milk in general is promoted as an all-around excellent food and there is no indication of geographically patterned variation in the ability to digest milk in adulthood. Importantly, milk is to be boiled before consumption—or warm from the source—but heating milk does not change the exposure to lactose. Only fermentation or separation of the curds from the lactose-containing whey results in reduced lactose in the resulting product. Consistent with the culinary history of dairy described in the previous chapter, cheese is not mentioned in the Ayurvedic literature as a dairy product—fresh curds, butter, *ghee,* whey, and buttermilk are the only milk products described.

Digesting Milk: Similarities and Differences between the United States and India

Discourse about the virtues of cows and milk in general, and cow versus buffalo milk in particular in India has a number of parallels in the United States. While the religious sentiments focused on the cow have no analogue in the United States, the cow is for all practical purposes the only source of dairy milk there, and hence any views about dairy are conditioned on the understanding that the milk comes from cows. However, over the past ten years or so, plant-based milk alternatives have become widely available and are marketed similarly to cow milk. Thus there is currently a comparative dialogue on the merits of different milks and it invokes concerns about digestion. There are also social movements that decry the treatment of cows in the production of not only beef, but also milk, which poses an interesting juxtaposition with Indian laments about the poor treatment of cows despite their exalted status.

Furthermore, while the cow doesn't achieve sanctity in the United States, and most Americans enthusiastically consume beef, cow milk is given a privileged status, and its presence in dietary recommendations is vigorously defended by government and dairy industry officials. Milk is, if you will, a "sacred cow" in this domain—its presence is largely uncontested, as are price supports for milk producers. The government buys up surplus milk and milk products in order to maintain a minimum price, and has done so since 1949. Milk has become part of a national narrative about what the ideal diet should be, and hence what model American citizens should look like. As part of this, substantial population variation in the ability to digest milk has been framed as a "problem," and ethnic minority Americans are counseled on ways to "overcome the barrier of lactose intolerance" (Jarvis and Miller, 2002) in order to conform to the idealized milk-drinking norm. However, milk has not always been viewed as an unequivocally good food for all, and the historical record suggests ways in which milk was understood to have particular effects on the body, especially digestion, and the ways that milk was "perfected" over the course of the past 150 years.

Divine Bovines in the United States

The cow may not have sacred status in the majority religions in the United States, but milk does have religious symbolism. In the book of Exodus in the Old Testament, Moses is charged to lead the Israelites to Canaan, described as "a land flowing with milk and honey"; in other words, the Promised Land, a land of great fertility where no one would want for lack of food and people would prosper. Milk and honey symbolized agricultural abundance and a thriving subsistence. Not surprisingly, this motif has been used as a brand name for a dairy: in marketing materials for the Texas-based Promised Land Dairy, where milk comes from prized Jersey cows, the "divine bovines" are featured roaming in the grass under a blue sky with puffy clouds (www.promisedlanddairy .com). While the divine bovine rhyme is tongue-in-cheek, the biblical quotation on the website seeks to establish a Christian rationale for cow milk, although the ancient Israelites herded goats and sheep along with cattle.

Such portrayals, along with those from the California Milk Advisory Board described in Chapter 2, promote an idealized image of cow care and devotion. However, there are quite public attempts to expose the dairy industry for its cruel or inhumane practices. People for the Ethical Treatment of Animals (PETA) is the most vocal of the animal protection groups active in the United States and is most concerned with the treatment of animals, especially in industrial production of milk and meat. PETA advocates against milk consumption because it supports harmful practices such as cow confinement and continuous pregnancies and milking. By extension, PETA highlights the benefits of a vegan diet and the potential negative health outcomes associated with dairy consumption.

PETA's interests would seem to be aligned with those of the *goshalas* in India, but PETA opposes milk production and consumption, while the *goshalas* actively protect cows and milk them. There is no shortage of lament in India about the poor living conditions of local cows, but a recommendation to abjure milk does not follow from this. Gandhi, who is upheld by PETA as a staunch defender of the principle of *ahimsa*, recognized the poor state of his country's cows, and decried their ill treatment: "How we bleed her to take the last drop of milk from her, how we starve her to emaciation, how we ill-treat the calves . . . how cruelly we beat the oxen . . . I do not know that the condition of cattle in any other part of the world is as bad as in unhappy India" (Gandhi and Kumarappa, 1954, 7). Despite such depictions, there is effectively no opposition to cow milk production in India, despite recognition of the often wretched state of its source. And the efforts to turn the populace away from buffalo milk stem not from the treatment of buffalos (which is often better than that of cows) but from pejorative views of the animals' personalities and the belief in the religious superiority of cows.

Digestive Concerns

As described in Chapter 2, cow milk was introduced into North America during the colonial period, as European settlers sought to re-create the diet familiar to them, and their views about milk were informed by a humoral theory of disease that predominated until the late eighteenth and early nineteenth centuries. Although these views were elaborated somewhat differently than in Ayurveda, there are clear connections between the two systems of medicine, likely reflecting a common Mediterranean source. The Roman physician Galen, following Aristotle's view, considered milk a "twice-cooked" form of blood (also referred to as "white blood"). A food with cooling and moist properties, it was considered to be especially well suited to the very young and old, whose humoral constitutions tended toward the cool, and who benefited from its ability to build flesh and blood (Valenze, 2011). As Deborah Valenze describes, "for consumers in mid-life, milk presented a slippery slope of nutritive seduction and betrayal. The viscous liquid most likely would spoil in the process of digestion, sending putrid fumes upward to the brain and chalky deposits downward to the kidneys, where it would create blockages. . . . Only the sturdy . . . could tolerate the digestive challenges of milk (2011, 61). Furthermore, the Italian Renaissance scholar Platina ranked goat and sheep milk above that of cows: goat milk was said to "loosen the bowels" (Valenze, 2011). Given that this was written in southern Europe, where the majority of the population loses the ability to digest milk in adulthood and goats and sheep were the main sources of milk, this may have been an accurate description. The similarities with Ayurvedic descriptions of milk, and buffalo milk in particular, are striking. While we might expect the descriptions of many foods to reference their effects on digestion, milk seems singled out for digestive challenges, particularly for adults.

But the United States was initially populated by peoples from northern, not southern Europe, whose regular diet included milk and (more commonly) a variety of dairy products. Cows were not the only source, however; mares, asses, goats, and sheep also provided milk, and individuals varied in which suited their humoral constitution: "asses milk is best, for some cow's milk, and for others goat's milk, because one cleanses, the other loosens, and the third strengthens more than the rest" (Moffet et al., 2011

[1746], 208). Still, it was recognized by Thomas Moffet, a British physician who wrote in the late seventeenth century, that cow milk was most widely used and, for otherwise healthy adults, who take it "now and then," it "nourishes plentifully, increases the brain, fattens the body, restores the flesh, assuages sharpness of urine, gives the face a lovely and good colour, increases lust . . . as for children and old men they may use it daily without offence, yea rather for their good and great benefit" (Moffet et al., 2011 [1746], 213). However, boiling of milk was not recommended, as it loses its "alimentary virtues," and young men should not drink milk as it would be bad for their teeth, "for there is no greater enemy then unto them than milk itself, which therefore nature has chiefly ordained for them, who never had or had lost their teeth" (Moffet et al., 2011 [1746], 211–212).

Butter and cheese were widely made dairy products, although their effects on digestion were variable: "new, sweet, and fresh cheese nourishes plentifully; middle aged cheese nourishes strongly; but old and dry cheese hurts dangerously, for it stays siege, stops the liver, engenders choler, melancholy, and the stone, lies long in the stomach undigested, procures thirst, makes a stinking breath" (Moffet et al., 2011 [1746], 219). Goat milk cheese was considered most nourishing, followed by that from cow milk, but ewe cheese was digested more quickly. And while the Dutch were considered to be especially partial to butter, as the following proverb attests, "Eat butter first, and eat it last, And live till a hundred years be past," the English were more skeptical. As with milk, butter was recommended for growing children and the aged, but considered unwholesome otherwise as it "hinders the stomach's closing, whereby concoction is foreslowed, and many ill accidents produced to the whole body" (Moffet et al., 2011 [1746], 218). This hydraulic conceptualization of the body and concerns about digestion and dairy consumption reflect in some ways similar views in Europe and India, but with variation in the details, especially vis-à-vis dairy products. The European version was transported across the Atlantic, where it remained an interpretative framework until it was gradually superseded by a biomedical one in the late eighteenth and early nineteenth centuries in the United States.

Digesting Milk: Biomedical Perspectives

Elaboration of the biomedical conceptualization of human physiology over the twentieth century altered how digestion was understood. Concurrent with developments in nutrition, digestive enzymes were identified as having specific roles in breaking down dietary components in the gastrointestinal tract, including lactase. It was assumed that lactase worked

similarly to other enzymes involved in the digestion of disaccharides insofar as it was "normal" to produce these throughout life in order to digest
the "normal" components of the diet. It wasn't until the second half of
the twentieth century that a new way of thinking about milk digestion
came about. Clinicians and nutritionists in the United States worked under the assumption that everyone could and should drink milk. Anyone
who experienced the symptoms of lactose intolerance—nausea, diarrhea,
gas, bloating, or general gastrointestinal discomfort after drinking milk—
was considered to have some underlying pathology. This view made sense
to people who were largely of European descent—after all, they had no
trouble digesting milk. This is an example of what I have termed "bio-
ethnocentrism," the belief in the superiority of a biology and set of cultural
practices, usually one's own, and the establishment of these as the norm
(Wiley, 2004, 2011c). In this case, the idea that everyone should drink milk
and have the digestive biology that allows for milk digestion without
unpleasant symptoms is an example of bio-ethnocentrism.

The formative studies demonstrating population variation in milk digestion came from researchers in Baltimore, Maryland, who found that
75 percent of healthy adult African American men but only 10 percent
of European American adult men (both groups were comprised of volun-
teers from a correctional facility) had "lactase deficiency" (Bayless and
Rosensweig, 1966). Furthermore, over 90 percent of the African Americans reported the symptoms of lactose intolerance, which is characterized by diarrhea, cramps, and gas after milk consumption, while only 10
percent of the European Americans did so. The symptoms of lactose intolerance stem from undigested lactose in the colon, which is then at least
partially fermented by bacteria (depending on the gut microbiota), with
gas as a by-product. Individuals with low levels of lactase who reported
symptoms of lactose intolerance had no other signs of intestinal pathology. Studies also confirmed that lactase activity in adults could not be
stimulated by providing lactose (Rosensweig, 1973). Subsequent studies
revealed a similar result: in most humans lactase production is turned off
during childhood (Bayless and Rosensweig, 1966, 1967).

These authors also speculated on the role of milk in the history of African populations from which African Americans are largely descended,
using the terminology of the time: "Milk drinking after weaning is very
unusual in the areas of central west Africa, whence the American Negro
came" (Bayless and Rosensweig, 1966). They were also quite concerned
that school lunch programs that provided milk to children were poten-
tially posing a problem for African American children (Paige et al.,
1972). Thus recognition that ethnic minorities might be adversely af-

fected by dietary policies that promoted milk consumption in the United States due to genetic differences in lactose digestion has existed since the early 1970s.

Concerns about lactose intolerance have grown since then, as has the understanding of population variation in lactase persistence. Within the United States, Native Americans, and those of African or East or Southeast Asian descent have very high frequencies of lactase impersistence (Wiley, 2011c). Those from southern Europe have higher frequencies than those from northern Europe. Given the ethnic diversity of the country and the widespread promotion of milk, it is no surprise that this has become a contentious issue in dietary guidelines and biomedical recommendations (Bertron et al., 1999). This is evident in the linguistic terms used to describe lactase impersistence: lactose *mal*digestion and lactase *deficiency* are commonly used, reflecting an implicit view that producing lactase throughout life, and hence digesting lactose throughout life is "normal" and "healthy," despite widespread acknowledgment that most adults in the world are in fact lactase impersistent. National Dairy Council researchers put it this way: "Data from most studies suggest that individuals with primary lactase *deficiency* consume less milk than those who digest milk *normally*" (Jarvis and Miller, 2002, 58; emphasis added). So while these researchers acknowledge underlying biological variation, it is not considered significant; it should not be a barrier to consuming milk and enjoying the health benefits it confers to those who have a history of drinking milk and continue to do so regularly. One can and should "overcome" this biological deficit to achieve full participation in U.S. culinary culture and its self-evident salutary consequences (Wiley, 2004). There has yet been no parallel discussion of lactose intolerance as a barrier to milk consumption in India, where poverty is a more profound obstacle. Milk intake in India is much lower on the whole, and the dairy industry cannot provide enough supply to meet demand, in stark contrast to the glut of milk in the United States, to which consumers are largely indifferent.

Perfecting Milk, Perfecting Citizens

Attempts to make *all* Americans into model milk-drinking citizens is a twentieth-century effort on the part of the dairy industry and nutritionists to get people to buy and consume more milk. They are predicated on the reliability and safety of a commercial milk supply. In the mid-nineteenth century, a growing urban populace had become disengaged from domestic dairy production, and relied on commercial sources for their milk. In this context, concerns began to be voiced about the quality of milk being sold and consumed. No longer directly involved with milking animals, consumers had to rely on those who were, or, more frequently, middlemen peddling in the cities, and concerns were raised about both milk adulteration and hygiene. It wasn't until pasteurization was widely adopted by an ever-larger-scale commercial dairy industry and growing consumer base in the early twentieth century that milk could (at least for urbanites) be considered an especially healthful food.

As Melanie DuPuis has argued, the past 200-year narrative about milk in the United States has been one that asserted milk's "perfection." "The perfect story portrays drinking this food fresh from the cow as an ancient custom and America as a milk-drinking nation from the beginning" (Du-Puis, 2002, 4), a statement that resonates with rhetoric about milk in India. While the historical record documents the origins of widespread fresh milk consumption in the late nineteenth century, both public health and dairy industry advocates created a story about milk that insisted on its universality and uniqueness as a "perfect food." A variety of social institutions and cultural values played a role in scripting this narrative: a growing dairy industry with close ties to government policymaking bodies, a culinary legacy that included use of dairy (though not necessarily fresh milk), religious groups bent on both social reform and finding justification for milk consumption in biblical sources, and racial biases toward the culture and biology of "white" Europeans that found a mirror in milk's "whiteness." Technological innovations provided the infrastructural necessities to support the story: pasteurization, canning, refrigerated transport and ultimately widespread home refrigeration, various mechanized means of increasing the efficiency of milking large numbers of cows, and the nascent field of nutrition research. Milk became both a fluid that was increasingly "perfected" in the modern period, but that also became a symbol for America's progress and march toward a "perfect union."

Milk's detractors, who gained some strength in the late twentieth century, have been cast as those who would impede the quest for perfection and contribute to the downfall of a civilization made mighty by milk (cf. Heaney, 2001). Robert Heaney describes groups such as PETA as "the most immediately dangerous," presumably because of their sometimes terrorist tactics (Heaney, 2001, 160). Further, their influence is decried: "We confront a recent, very modern efflorescence of militant groups that oppose all use of animal products and aim to effect a nutritional policy outcome similar to that of the creationists with regard to evolution. Those who care about nutrition, those who think nutrition important for the public health general, need to realize that the present-day skirmishes may be only the first wave of a growing battle" (Heaney, 2001, 163).

The story of perfect milk came to be dominated by nutritionist views, which came to the fore in the early twentieth century. By then the ways in which milk or dairy products in general were conceived as healthful began to shift as a function of the discovery of vitamins (i.e., those food constituents "vital" to life), such as vitamin A and the B vitamins. Elmer McCollum termed this "the Newer Knowledge of Nutrition," and his and others' work established a new rationale for the "vitality" of milk (McCollum 1922, 1957). Later, researchers placed minerals such as calcium at the forefront of milk's nutrients. Robert Heaney, who is quoted above, is an osteoporosis researcher, and this disease is portrayed as the scourge that can only be avoided with regular consumption of milk, primarily due to its high calcium density and its fortification with vitamin D.

Nutritionist messages about milk abound in the United States, and are circulating widely in India and many other countries. Milk is hailed as a source of energy, protein, and, most especially, calcium. Without competing concerns about the source of milk or about the religious significance of cows, milk's meaning has been uniquely elaborated within a nutritionist context. But it is important to keep in mind that these nutritionist understandings are underpinned by assumptions and institutions that uphold a particular national narrative about the importance of milk, framed in relation to the "fitness" of the nation and its citizens. Much like Gandhi's and contemporary Hindu politicians' advocacy of cow milk, there is a nationalist thrust to milk promotion efforts in the United States. For example, in their ode to the virtues of milk, *The Most Nearly Perfect Food: The Story of Milk*, Samuel Crumbine and James Tobey wrote:

> Throughout the course of history, milk has been hailed as the ideal food, an opinion which is well justified by our modern knowledge of the science of nutrition. The races which have always subsisted on liberal milk diets are the ones who have made history and who have contributed the most to the advancement of civilization. As was well said by Herbert Hoover in an address on the milk industry delivered before the World's Dairy Congress in 1923, "Upon this industry, more than any other of the food industries, depends not alone the problem of public health, but there depends upon it the very growth and virility of the white races." (1929, 77–78)

The twentieth-century nationalist rhetoric about milk's contributions to the political and economic success of the United States—and the "virility of the white races"—conveniently ignored the long-standing dairy traditions of India. Moreover, since it was milk, rather than cheese or other dairy products to which these triumphs were attributed, this is especially ironic, since India has a longer and more extensive milk-drinking (as opposed to cheese-making) tradition. In Crumbine and Tobey's global history of milk, India is mentioned as a place where pastoralist peoples of the remote Himalaya had a "magnificent physique" attributable to their diet of goat milk and vegetables. But only those Indians living in "the few good dairy regions of that country are always vastly superior to the more numerous natives who live only on cereal grains" (1929, 9–10). Thus there would be no "non-white" large-scale dairy-based civilization to compete with the Euro-American mythology, which causally linked Western political and economic success and dairy culture. Only isolated pastoralist groups, living in an Eden-like state, could be credited with achieving similar vigor—and given their small numbers, they did not pose a threat to the narrative of Western supremacy. The story also points to the long-standing belief in milk's unique ability to create "superior" bodies, which merges nicely with nationalist goals, the topic of Chapter 5.

Multiple Milks in the United States

Until recently, viable alternatives to milk consumption were not available in the United States, and certainly not highlighted in dietary recommendations for Americans. Strong resistance to those came from various dairy groups, including the National Dairy Council. Food groups and the more recent U.S. food pyramid put dairy products in a separate category, reflecting the unique status of this single food, just as was laid out in the Ayurvedic *Charaka Samhita*. Indeed, in the current MyPlate (Figure 2.5), the visual translations of the Dietary Guidelines for Americans, 2010, "Dairy" retains its separateness, as it is featured in a glass, set off from the plate. It is also indicative of milk intended to be served cold—although this only became possible with widespread usage of refrigeration and it was contraindicated in Ayurveda. But, in the USDA textual guidelines it reads: "Increase intake of fat-free or low-fat milk and milk products, such as milk, yogurt, cheese, or fortified soy beverages (U.S. Department of Health and Human Services and U.S. Department of Agriculture, 2010, 34). Furthermore, "Soy beverages fortified with calcium and vitamins A and D are considered part of the milk and milk products group because they are similar to milk both nutritionally and in their use in meals" (U.S. Department of Health and Human Services and U.S. Department of Agriculture, 2010, 38). Thus nonmammalian milk has emerged as a viable alternative to cow milk, largely to avoid the digestive difficulties that many U.S. citizens encounter with dairy milk, and they are positioned as "drinks," not as part of the food on the plate.

A cynical view of the revised recommendations is that they only occurred because the largest soy milk producer in the United States, Dean

Foods, is also the largest dairy processing company in the country, with net sales of over 13 billion dollars in 2011.[1] Thus, the inclusion of non-dairy milks is no longer a contentious issue for this conglomerate, although it remains so for other major dairy interest groups such as the National Milk Producers Federation, who contested the use of the term soy "milk" in a letter to the U.S. Food and Drug Administration (FDA) on the grounds that it does not qualify as milk (i.e., a mammalian lacteal secretion).[2] Notably, other nonmammalian milks beyond "fortified soy beverages," such as almond, coconut, or rice milk are not mentioned in the dietary guidelines as suitable alternatives. While these generally are fortified with calcium and vitamin D to provide the nutrients most hailed as the important nutrients in dairy milk, only soy milk has a similar amount of protein, which is probably why it is specifically in the dietary guidelines. Beyond the fact that these milk alternatives contain no lactose, they have other appeal to consumers as well, especially those concerned about the human biological effects of recombinant bovine growth hormone (rBGH or rBST), which is used by some dairy farmers to stimulate greater milk production in cows.

Although plant-based milks are relatively new as mainstream alternatives to dairy milk, at least in their current formulation—that is, packaged, fortified, and sold like dairy milk (and found in the dairy case of supermarkets)—they have been known for centuries in Europe. Almond milk has been made in southern Europe, while soy milk has been known in East Asia as a more fluid form of tofu. But in the United States they are novel, and thus far they are not a major competitor to dairy in India. Perhaps an analogy can be drawn between cow milk in the United States and buffalo milk in India: both require a "strong digestive fire," with alternatives being easier to digest. But plant-based milks are unlikely to catch on as nationalist drinks in either country: cow milk is a viable alternative to buffalo milk in India, while in the United States dairy's meanings are quite different from those imparted to plant milks. It does remain to be seen if plant-based milks ever catch on in India. In light of the discussion of PETA above, and known geographic variability in lactase persistence, it would seem a possibility, but in both places strong cultural biases toward cow milk will likely preclude widespread adoption.

Shared concerns about the digestion of milk exist in India and the United States, and have their roots in humoral understandings of the body and the evolutionary histories of the peoples inhabiting both countries. Debates about what kind of milk to drink are alive and well, and are closely related to effects these different milks have on the body, above and beyond the digestive system. They are heard throughout the social and political institutions that influence policymaking at the state and national levels, whether reflected in the USDA/DHHS Dietary Guidelines for Americans, Gandhian philosophical musings used to establish a newly independent India, or contemporary Hindu nationalist political posturing. In the next chapter I turn to specific meanings attributed to bovine milks in relation to building the bodies of Indian and U.S. citizens.

5

Milk as a Children's Food
Growth and the Meanings of Milk for Children

The idea that children *should* drink milk makes intuitive sense—the first and only food of infants is mother's milk, and so substituting other milk for breast milk, augmenting the diet of a nursing infant, or transitioning a baby or toddler to another animal's milk after weaning seems like a reasonable way to approach the nourishment of young children. Milk from other mammals such as cows, water buffalo, or goats and sheep has long been considered a particularly appropriate food for children, at least among populations with traditions of keeping these animals. Feeding the milk of another maternal mammal is much like employing wet nurses or having other lactating women nurse a child. However, the role milk has played in the diets of older children, or at what age milk was no longer seen as uniquely valuable in the diet, is an open question.

Yet, as outlined in the first chapter, human and bovine milks are not equivalent. Indeed, bovine milks are generally too low in iron for infants, and reliance on those milks as a primary food can result in severe anemia and intestinal bleeding (Ziegler, 2007), and protein and calcium content is also much higher. Dilution of milk (either fresh or powdered milk) with water ameliorates the latter problems, but can increase the risk of microbial contamination. Households that do not have access to their own dairy animals must rely on market sources, and these may be highly suspect as milk may be adulterated, contaminated, or otherwise dangerous to infant health. Even those with their own source face problems with infection (depending on the hygiene of the animal and its handler) and the nutritional challenges that milk presents to children. While it may be viewed as appropriate for children to drink milk, there is some ambivalence about the use of nonhuman milk for infants, due to concerns about milk's safety, composition, and reliability.

Apprehensions about the nutritional qualities of nonhuman milk are mitigated as infants mature and start eating a more mixed diet, although problems of contamination and adulteration remain. In the United States, these worries have largely disappeared with pasteurization and regulation of the commercial milk supply, and as a public health infrastructure that ensured clean water developed in the early twentieth century. These efforts exist in India as well, but unease about milk adulteration is widespread among those who buy milk from the market or the local door-to-door milk seller, and milk is often diluted with water, the sanitation of which varies across locales. Thus while feeding infants and children mammalian milks may "make sense," it has been accompanied by some anxiety as well, especially when milk is not available from trusted sources.

This chapter is concerned with how milk (cow, water buffalo, or goat) is positioned as a particularly appropriate food for children, highlighting those beyond weaning age. Ideas about milk as particularly beneficial to children are well entrenched in both the United States and India, and supported by a variety of government initiatives, including dietary guidance and, in the United States, school feeding programs that extend milk's usage among older children. But overall milk consumption among children is declining in the United States, while it is on the rise in India, although we have less information on what age groups are drinking more milk there. Thus milk's meaning in relation to children's diets may be quite different in each place, but I argue that both trends center at least in part around the assumption that milk enhances children's growth, especially in height. I trace the history of ideas about milk as a food for children in both countries, and consider how ideas about milk's contributions to growth came to the fore in the twentieth century. This also requires a look at ideals and meanings surrounding variation in body size. Scientific studies of milk's contributions to growth will also be discussed briefly.

Historical Context: Milk and the Feeding of Young Children

Among populations with dairy animals, the use of milk to supplement breastfeeding, or as a weaning food, probably has a history as old as domestication itself. This is conjecture, as we have no clear evidence one way or another, but milking seems to have been one of the earliest uses of domesticated bovines in the European context (Craig et al., 2005). Currently among countries in Africa and Asia, mammalian milk is the second most common fluid (water being the first) given to infants less than one year of age (~12% of zero- to six-month-olds and 30% of six- to twelve-month-olds received milk). In a national survey in India, 21 percent of mothers reported giving their zero- to six-month-old infants milk, and 45 percent gave their six- to twelve-month-olds milk while they continued to breastfeed (Marriott et al., 2007). Thus using mammalian milk as a substitute for or supplement to breast milk for infants is a well-known practice.

Both in ancient Egypt and Imperial Rome, infants appear to have been fed milk from local domesticates starting from around six months of age. According to osteological evidence, their health and growth did not seem to have suffered with this diet, and such a practice was recommended by medical writers of the time (Dupras et al., 2001). In India, Achaya (1994b) reports a similar pattern for the Vedic period—infants were given their first solid food of boiled rice, milk, sugar, and honey, and perhaps even some meat, starting at around six months of age. Honey was similarly given in Egypt, and this remained, along with giving infants wine, a practice throughout the medieval period in Europe.

The classic Ayurvedic texts of Susruta described in Chapter 4 have an elaborated and systematic discussion of pediatrics that, not surprisingly given the overall Ayurvedic concern with diet, includes recommendations for child feeding. Honey and clarified butter were advised within the first few days of life. The colostrum was discarded and mature milk was considered to begin on the fourth or fifth day. Breast milk was considered "the natural food of an infant" but if it was not available, either from the mother or wet nurse, cow or goat milk could be an adequate substitute, as both were considered "efficient vitalizing agents" (Kutumbiah, 1959). Children were divided into age groups by their normative diet: milk alone (meaning breast milk, or cow/goat milk if substituted; up to one year); milk and boiled rice (one to two years); boiled rice alone (older than two years) (Bhishagratna, 1911, vol. 1, 322). The process of growth was thought to extend up to twenty years, but there is no recommendation for milk beyond two years of age or an assignment of an appropriate time to wean from breast to other milk. Milk seemed to be associated with young children's diets up to age two years, at which point they would be moved onto the adult diet, which may have included milk but was not characterized by a mandate for it. Thus children beyond weaning age (~two years of age) were not singled out for milk consumption above and beyond what adults might have been consuming.

Currently the World Health Organization (WHO) sets widely adopted standards for infant and young child feeding. Following WHO guidelines, the Indian National Institute of Nutrition recommends exclusive breastfeeding for the first six months, which should continue for up to two years (National Institute of Nutrition, 2010). If breastfeeding is not possible, boiled milk from cows, goats, or buffalo can be substituted, diluted to reduce the protein content, and sugar added, even for newborns. At four weeks, no dilution is required, although sugar should be added, and infants supplemented with both vitamin C and iron, which are limited in bovine milk. Specialized formulas are not recommended.

This is in contrast to recommendations in the United States: while exclusive breastfeeding is recommended through six months (as per WHO guidelines), if breastfeeding does not occur, infants should be fed specialized infant formula, which is usually based on fortified cow milk. However, whole cow milk should not be fed until after one year, due to its high protein but low iron content (American Academy of Pediatrics Committee on Nutrition, 1992). Use of infant formula is widespread, with children's food marketers attempting to extend their use well after the first year (e.g., "toddler formulas").

Children's Milk Consumption in the Post-Weaning Period

Beyond these prescriptive materials, we have little idea about milk in the diets of children in Europe and the United States until the nineteenth century or in India until the twentieth century. There is scant characterization of children's diets after weaning in European history (Prowse et al., 2010). Indications are that children would be recipients of milk if it was available above and beyond that used for making cheese or butter (in Europe/ the United States) or curd and *ghee* (India), but that, by and large, children consumed the same foods as adults after they were weaned (cf. selections in Lysaght, 1992). Indeed, ideas that post-weaning-age children had "special" dietary needs that reflected their biological distinctiveness compared to adults (their ability to grow in height notwithstanding), are relatively modern, marked by the emergence of pediatrics as a recognized medical specialty within biomedicine in the 1920s. Pediatricians were beginning to articulate the ways in which children faced health issues that were quite different from those of concern to adults, and there was greater professional oversight of child health. Pediatricians were eager to assert their expertise in the area of child feeding, especially infant feeding, and were centrally involved in the development of "percentage feeding," the modification of cow milk constituents to better mimic breast milk (Halpern, 1988). In addition, a more sentimentalized view of childhood was emerging, at least among the middle class (Ariès, 1962; Calvert, 2003). Harvey Levenstein and other historians noted a trend toward a more "child-centered" society in the United States in the 1920s, spurred on by anti–child labor laws and compulsory education, which also served to lengthen the life stage recognized as childhood (Calvert, 2003; Cunningham, 1998; Levenstein, 1988).

The idea that older, school-age children should drink milk, or be privileged consumers of it, is a relatively recent one in U.S. history, and even newer in India, to the extent that it is an idea held by at least some portion of the population. Focusing first on the United States, this idea has been traced to the mid- to late nineteenth century, and dates to the very origins of widespread fresh milk consumption, as described in Chapter 2. Infants were the first market for fresh milk, but as the volume of milk production increased, distribution networks developed, and methods of preservation (canning condensed and evaporated milk; refrigerated transport) were more widely adopted, growing the market for milk became a goal of the emerging U.S. dairy industry. The target market identified was older children, who were, after all, still growing, albeit at a slower rate than infants. They represented a natural extension of "the milk years." The National Dairy Council (NDC) became the primary institution promoting milk to children, and set about creating educational campaigns to extol the benefits of milk consumption to child growth. Child health came to be equated with milk drinking, and through NDC's efforts, combined with government programs, education campaigns, and clinical recommendations, it became increasingly difficult to imagine how children of all ages could be healthy without drinking milk.

While nineteenth-century public health reformers brought the contamination of the milk supply and its deadly effects on children to public attention, they worked from the assumption that children should be drinking milk and that cow milk was "naturally" a food needed by them. Milk's white color, maternal associations, and Old Testament odes to a land that "floweth with milk and honey" were deployed in support of this view. Robert Hartley, the evangelical social reformist who railed against the swill milk system, predicated his crusade for safe milk on the assumption that cow milk was a natural drink for children: milk is "the chief aliment of children of all ages in all places where the population is condensed in great numbers; it is the nourishment chosen and relied upon to develop the physical powers and impart vigour to the constitution during the most feeble and critical period of human life . . . it [swill milk] is the pernicious sustenance upon which we depend as the staple diet for children" (Hartley, 1977 [1842], 109, 136). In his book on the milk trade in New York, John Mullaly similarly described milk as a "principle article of food of all children " (Mullaly, 1853, 23).

Figure 5.1 Horlick's malted milk as a food for schoolchildren, using the theme of Shakespeare's Seven Ages of Man "The schoolboy with his shining face." *Ladies Home Journal*, June 1904, p. 57.

As noted above, it remains unclear where this assumption came from. It may have long been latent, part of an intuitive recognition that since infants begin life consuming milk, milk provides resources to support growth that could easily extend to post-weaning-age children. Furthermore, since infants need milk every day, perhaps older children should also consume it that frequently. Whatever its origin, the dialogue about children and milk was (conveniently, for the dairy industry) framed not in terms of whether children (that is, post-weaning-age children) needed milk, but rather in terms of the importance of hygienic production.

Given that fresh milk still had a high likelihood of contamination well into the twentieth century, it was initially difficult to promote milk as an especially salubrious beverage, especially for children. Indeed the earliest advertisements were for tinned, rather than fresh milk, and the safety of the products was highlighted. Healthy plump babies who were "strong" and "contented" graced the labels of tinned milk. But after the turn of the century and throughout the twentieth century, labels and promotions increasingly tended to feature older children, from toddlers to teenagers, and more often fresh milk rather than tinned milk.

Most illustrative are a series of advertisements for Horlick's malted milk, which utilized Shakespeare's Seven Stages of Man from the opening lines of *As You Like It:* "And then the whining school-boy, with his satchel. And shining morning face, creeping like snail. Unwillingly to school" (see Figure 5.1). (Notably, the "whining," "creeping," and unwillingness to go to school are omitted in the ad). This was an explicit attempt to promote a processed form of milk (malted, powdered, sweetened) not just to babies, but to individuals throughout life. And in its powdered form, milk's status as a food-drug hybrid is clear—its health enhancing, or ill-health correcting abilities are featured. But further, this processed form served to make milk more palatable for children, and insofar as children's foods came to be highly sweetened and/or colorful and more highly processed versions of adult foods throughout the twentieth century, milk in this form was a harbinger of that trend. Chocolate milk was introduced in the 1920s to get "Little Milk Rebels" drinking their milk (Levenstein, 1988).[1]

Moral Tales of Milk

The lack of emphasis on milk as a food for post-weaning-age children before the twentieth century is reflected in the ways in which milk was featured in popular folktales, or those designed to convey messages about proper socialization and behavior. Given the importance of milk to their

economies, milk was definitely part of the subsistence milieu in both
northern European and Indian folktales, but not as a uniquely children's
food. In light of the normative status that children's milk drinking was to
attain (i.e., that children *should* drink milk), it is notable that milk has no
such moral status in fairy tales or children's stories on either continent.
Thus when contemporary parents reach for a convenient parable for why
their children should drink milk, none are readily at hand. Although milk
and bread—or in the Indian case, milk and rice—are ubiquitous as children's food, there are no cautionary tales about the perils of not drinking
milk—children who never attained their full height or who wasted away
from lack of milk.

In fairy tales, milk symbolizes abundance and riches (turned into *ghee*
it can be sold for a good price), a life of ease (so much milk, it can be used
in a luxurious bath), and purity and wholesomeness (in Hansel and Gretel,
the evil woman serves the children "good food"—milk, pancakes, sugar,
apples, nuts—these foods being a perfect foil for her malicious intent). In
Indian folktales, milk is strengthening and refreshing, but frequently
served to husbands, who rank more highly than women and children in
the household hierarchy. Giving someone milk (whether it be cow milk
or buffalo milk, or a woman's own milk) is portrayed as nurturing, but
only in the case of breastfeeding is it uniformly directed at the young.
Neither the European tales of the Brothers Grimm and Hans Christen
Andersen, nor the Panchatantra, Hitopadesha (both written in Sanskrit),
or Jataka (written in Pali), all ancient moral tales about proper behavior
and guidance to live a good and healthy life that date back to at least
300–200 BCE in India, offer an edifying moral about the ill consequences
of children not drinking their milk.

Milk serves a common role in the folktales of India and Europe as an
offering to strangers, visitors, husbands, or to animals, who may have
magical powers or be someone else in disguise to "test" a person. Offerings
are indicative of generosity, humility, and honesty. Furthermore, cows
mark wealth as the producers of milk, and milk is a precious commodity.
Cows who do not give milk signal impoverishment or ill treatment. Tam-
pering with milk is a particularly egregious break in the social contract.
One Indian tale, "Dishonesty of Sriniwas," describes an honest milkman
who is convinced to increase his profits by diluting his milk with water, a
common practice among milk sellers. Not surprisingly, as he sits to rest
by a river one day, half of his bag of coins disappears into the river. He realizes his folly—"earning of water had sunk into the water." A story about
a widely acknowledged fraudulent behavior involving milk makes its
point about honesty and integrity.

More famous in India are the tales of Krishna as a young child, whose mischievous ways centered around stealing and eating butter. Krishna is featured as a plump baby with a glint in his eyes and his hand in the butter vessel (see Figure 5.2). Tales of his childhood are replete with ideals for motherhood—unconditional love, nurturance, and indulgence (tempered by occasional irritation!) all situated among an abundance of dairy products. Krishna's fatness is symbolic of his penchant for all things

Figure 5.2   The young Krishna stealing butter. Image ca. 1750.

dairy—especially butter—but also his mother's devotion. Subsequently his youth was spent in an idyllic state of frolicking with the cows and cowherds in bucolic pastures—the very essence of a carefree life and abundant food—before an adulthood spent vanquishing various evils and defending *dharma*. Although Krishna is associated with cows, milk, and butter, stories about him carry no moral imperative for children to drink milk, and he is not held up as particularly tall, large, or strong (his supernatural powers underlie the latter), and his plumpness is limited to infant representations.[2] On the other hand, one could argue that the status of milk as a "good" food is reinforced by its association with Krishna, who stands on the side of the virtuous, and whose body may represent an ideal masculine form.

A more recent short story from Bengal reveals how milk is at the center of family conflicts in households with limited income. "A Drop of Milk" by Narendra Mitra (2007) describes a poor family who can afford milk only to meet the 1.5 pints required to fulfill the needs of their one-and-a-half year-old daughter. The father gets only powdered milk for his two cups of evening tea, which he finds disgusting. He implores his wife to indulge him in a bit of "real" milk for his tea. Noticing how haggard he had become from working so hard, she spends some of their limited funds and serves him not just tea with milk, but a whole cup of milk. Embarrassment ensues for the father as he drinks the milk in front of his nine-year-old son doing his schoolwork, but he is not sufficiently chagrined to give up his cherished milk, and subsequently his health improves. But soon everyone has a claim on milk—the mother for her digestive "acidity," the son for his schoolwork, the brother-in-law to get a job, and the father is no longer getting his treasured daily cup. The story ends with each household member acknowledging the greater need of the other and ultimately directing the contentious glass of milk to an undernourished schoolmate of the son.

While the infant daughter's need for three cups of milk is uncontested, other claims on this valuable food are up for negotiation—does the father have a clear right to the milk, given that he is the sole income earner and working very hard, or is it the son who needs it for success at school, the brother-in-law who needs to get a job, or the wife who is suffering from indigestion? Consistent with Ayurvedic principles, the toddler has the first claim, but beyond that there is no clear privileging of the school-age son. Children's needs compete with those of other household members, although there is recognition in the father's embarrassment over drinking milk in front of his son that maybe the school-age boy should really have priority over the father. This situation reflects the importance of the household as a corporate unit in India (cf. Seymour, 1999), compared to the United States, where there is a more atomistic approach to individual roles in the household and a clear mandate that children—regardless of age—should get milk.

Selling Milk through Growth in the United States

It was through milk's association with child growth that milk was to become entrenched in children's diets in the United States, and this linkage has had broad salience in India as it moved toward and then beyond Independence, and became a central player in the global economy in the 1990s after a half-century of protectionist trade policies. As knowledge about nutrition and the effects of nutrients on human biology grew in the 1910s and 1920s, there were parallel public health campaigns focused on child growth. Milk advertisements fit right in with these. Among early twentieth-century ones were statements that milk "builds up the body—brains—bones—and muscles—and promotes growth of the entire system." A milk-growth linkage was exploited extensively by the NDC and other institutions promoting milk, initially basing their claims on Elmer McCollum and others' studies demonstrating that milk was rich in a variety of nutrients, and that these made rats and mice supplemented with milk grow at faster rates and to larger sizes than those that were not supplemented (see McCollum, 1957).

A series of studies on humans solidified this linkage. The most widely cited were two seven-month-long studies of children in urban working-class households in Scotland and Northern Ireland in 1926–1928 (Leighton and Clark, 1929; Orr, 1928). Three cohorts of school children ages five to six, eight to nine, and twelve to thirteen were given 0.75 to 1.25 pints of milk each school day. They were measured at the beginning and at the end of the two seven-month periods. The results (displayed in Figures 5.3) show that there were consistent differences between the milk-supplemented groups and both the control group and those getting the biscuit. The most useful comparison is between the biscuit- and skimmed-milk supplement groups, as these two groups received the same number of calories (it is assumed that these additional calories did not supplant other foods in the diet). The children who received skimmed milk grew 0.20 to 0.42 inches (0.51 to 1.07 cm) more than the biscuit-supplemented children over the seven-month period, raising the question of whether differences in growth would be even more accentuated if the additional milk had been consumed for longer periods of time, or throughout childhood.[3]

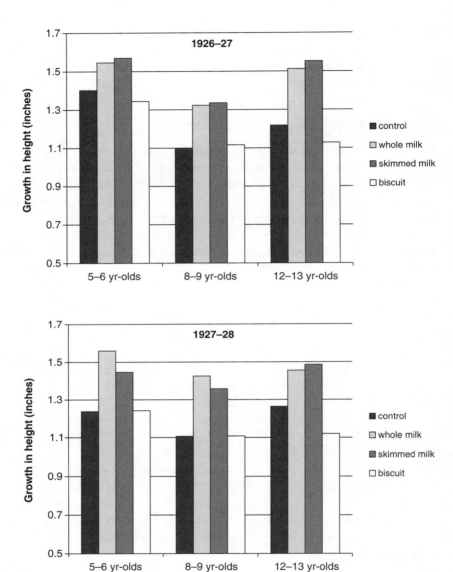

Figure 5.3 Results from two seven-month studies of the effects of milk supplementation on growth in height among urban working-class children in the United Kingdom. Based on data presented in Orr, J. B. 1928. Milk consumption and the growth of school-children. *Lancet,* i, 202–203; and Leighton, G., and Clark, M. L. 1929. Milk consumption and the growth of schoolchildren, *Lancet,* i, 40–43.

These studies provided the evidence to back up claims about milk's "special" effects on growth in height and set the stage for a variety of public health, pediatric, and commercial endeavors. Pediatricians, parents, and milk advertisements quickly latched on to the results and milk was then positioned to be the go-to food for children. Pediatricians institutionalized the practice of "well-child visits" that had questions about child feeding practices and physical measurements (especially weight and height) as their main emphasis. Working in concert with municipal reformers interested in reducing infant and child mortality, they established urban child-welfare centers, where children (both sick and healthy) would be measured, and mothers instructed in proper child-rearing techniques to achieve model behavioral and growth outcomes. Although the original purpose of such centers had been to provide sanitized milk for infants, under their broader mandate they included the dispensation of milk for older children and provided regular growth monitoring of children (Halpern, 1988). The U.S. Children's Bureau provided educational materials, which included pamphlets that emphasized the value of milk to child growth (Reaney, 1922). Thus through the confluence of government and philanthropic organizations, the professionalization of pediatrics, and a well-organized dairy industry, child health, growth, and milk became one package, one that upwardly mobile families were eager to embrace.

School health programs also took up the charge to encourage milk consumption among school children. In various localities special school milk programs had been established in the 1920s. Students generally paid for these supplementary meals, but they were often subsidized by local charitable organizations such as the Red Cross or Anti-Tuberculosis Association for needy children (Crumbine and Tobey, 1929). Children identified as underweight were eligible for additional milk. Moreover, children were often served milk in the classroom "while the teacher tells them about this wholesome product . . . [and] . . . to cultivate a taste for milk as the one best food for growth" (Crumbine and Tobey, 1929, 155). The Bureau of Education in 1922 offered curricular recommendations to promote milk drinking in schools, since it appeared that half of a large sample of American school children were not drinking their milk daily (Reaney, 1922). About one-third of them were drinking coffee while 20 percent drank tea. The report argued that "American parents and children must be taught the nutritional value of milk . . . because milk is the best and most important food in the diet of the school child. No other food can take its place. It contains the elements necessary for the growth of the different structures of the body" (Reaney, 1922, 4, 8). To this end a variety of educational activities were proposed in which milk was featured in every subject. Students charted their daily milk consumption while their weight and height were tracked across the school year. Similar types of activities can be found in the educational materials that the National Dairy Council makes freely available to schoolteachers in primary and secondary schools (cf. http://school.fueluptoplay60.com).

As noted in Chapter 2, school milk programs became federally supported in 1946 with the National School Lunch Act: "It is hereby declared to be the policy of Congress, as a measure of national security, to safeguard the health and well-being of the Nation's children and to encourage the domestic consumption of nutritious agricultural commodities and other food, by assisting the States, through grants-in aid and other means, in providing an adequate supply of food and other facilities for the establishment, maintenance, operation and expansion of nonprofit school lunch programs" (Gunderson, 1971). Fluid milk must be offered as part of school lunch programs in order to receive federal reimbursement and, 20 years after the School Lunch Program was authorized, private institutions devoted to the care and training of children also became eligible for milk reimbursements. Such efforts expanded in the early 1970s with the Special Supplemental Program for Women, Infants and Children (commonly known as WIC), which provides vouchers for a set list of foods for low-income pregnant women and new mothers and their children up to age five; children ages one through four years are allocated 16 quarts of milk each month (15.1 liters; roughly 1 pint or 500 ml/day). Through these federal mandates, milk became an essential part of school and child feeding programs, and links between the USDA, the NDC, and the professional organization of school meal administrators, the School Nutrition Association, are tightly interwoven. But given that support of school feeding programs explicitly exists to advance national agricultural interests, it comes as no surprise that there has been robust critique of these programs as a means to dispose of agricultural surpluses (including milk) and control the price of milk (Lanou, 2009; Nestle, 2002; Yeoman, 2003).

The U.S. government's rationale for subsidizing milk was that it would "safeguard the health and well-being" of American children, as assessed by height and weight gains. Thus the connection between milk and growth, be it of children or the agricultural economy, was solidified in federal agricultural policy. Not surprisingly, this link was thought to occur through milk's rich nutrients. Initially E. V. McCollum proposed that vitamin A was milk's unique growth-promoting nutrient. As the twentieth century wore on and more research came to light on micronutrients (vitamins and minerals), calcium came to predominate as the most important of milk's nutrients that might otherwise be limited in the American diet and that would appear to play a critical role in growth, given that calcium is an essential component of the skeleton.

Advertisements highlighting milk's calcium started in the 1930s, and became more frequent in the 1960s. Calcium remains milk's most boasted-about nutrient, symbolizing the inherent goodness and superiority of milk as a food, but it did not have this status in the early years when milk was being established an essential food for children. The idea that calcium contributes to "strong bones" only became well entrenched in the latter part of the twentieth century. This is a key message in nutritional education campaigns (most sponsored by the National Dairy Council, but with strong support from government agencies concerned with health and nutrition as well, cf. National Institute of Child Health & Human Development, n.d.). Calcium was what made milk "special." While other foods certainly contain calcium (especially dark green leafy vegetables), these sources have been disparaged by researchers supported by the dairy industry as having structural components that reduce the bioavailability of calcium (Heaney and Weaver, 1990; Weaver et al., 1997). Cow milk contains ~300 mg of calcium per cup (~120 mg/100 ml), which makes up one-fourth to one-third of the Institute of Medicine's Recommended Daily Allowance for calcium (this ranges from 700 mg for children under three years to 1,300 mg for adolescents, Ross et al., 2011). Cow milk is rich in calcium to support the growth of a large bovine skeleton; human breast milk contains only about 80 mg per cup (32 mg/100 ml), reflecting the skeletal growth needs of a much smaller infant. Milk advertisements trumpet calcium and all of the benefits (only some of which are bone related) that stem from it; but at the same time, the suite of milk nutrients is also highlighted. That is, milk has a "unique nutrient combination" that sets it apart from other foods or even food groups (cf. http://www.nationaldairycouncil .org/HealthandWellness/Pages/DairysUniqueNutrientCombination .aspx). So, milk's "specialness" in relation to growth has primarily been articulated through its nutrient density and unique combination of nutrients (although the same could be said about any given plant or animal food), and this nutritionist portrayal is well understood by most Americans. It is the primary rationale for dairy having its own "food group" in dietary guidelines.

Milk Scarcity and Food Assistance Programs in India

At the same time as the dairy industry was consolidating its relationship to child health and growth in the United States (and in Europe), British families in India at the height of the Raj in the late nineteenth and early twentieth centuries attempted to stay current with dietary trends at home, and feeding their children milk was one way of enacting this. But, without the dairy industry infrastructure that was becoming elaborated in the United Kingdom, households met their milk "needs" by keeping a cow. Concerns about hygiene were acute, reflecting both fear of contamination by local Indians, as well as fears of adulteration and illness from milk that were widespread in urban British centers at the time. Mothers were very careful about their children's food, especially their milk. Often a special nursery cow was kept, and the cow keeper had to bring it around to the verandah where the *amah* (nursemaid) was supposed to inspect his hands for cleanliness before watching him milk the cow, and to ensure no water was mixed in (Burton, 1993). The milk then had to be boiled thoroughly, and when the children went off to tea parties, they would be given their own bottles of milk carefully wrapped in tissue paper to take along, since no other mothers were trusted to boil the milk properly. Thus milk fears were not confined to concerns about native contamination. But despite these concerns, milk was believed to be essential to children's diets and much effort was put into ensuring its safety.

British colonial administrators were interested in the dairy industry in India, but prior to 1900 expressed no real concern with nutritional conditions among the native population (Arnold, 2000). It wasn't until the 1930s that nutritional research came to the fore as a priority for funding. Under W. R. Aykroyd's leadership of the Nutrition Research Laboratories established at Coonoor in South India, investigations into dietary intake and nutritional status were undertaken. Aykroyd had been involved in trials of milk and growth in the United Kingdom, and served as the first nutritionist of the League of Nations when it was established in the wake of World War I (Carpenter, 2007). He came to his work in India well versed in the current knowledge of micronutrient deficiencies as well the work on milk and growth, and was charged with ascertaining nutritional status among Indians and developing some affordable means of improving it. Not surprisingly, given his background and the existence of local dairy traditions, this phase of his work focused on milk. But he also had a broader public health message, arguing that undernutrition was routine, and not confined to famine periods, and that the state needed to become actively involved in the production and distribution of food (Arnold, 2000).

Aykroyd and his colleagues conducted surveys of child height and weight in three towns in South India (Coonoor, Mettupalayam, and Calicut) that were published in 1936 (Aykroyd and Rajagopal, 1936a). These can be usefully compared with a sample of U.S. children measured by researchers at the University of Chicago during the same period, which was during the Great Depression (Richey, 1937, as shown in Figure 5.4). However, the U.S. sample was drawn from private school students, and as such represents a relatively well-off segment of the population. As is clear, the American children were 3.5 to 7.5 inches taller than the Indian children of the same age, and while the absolute differences became more accentuated with age, as a percentage, the relationship did not change dramatically with age. Indian boys attained between 88 and 92 percent of the height of American boys from the same period. Interestingly, a comparison of a 2008 survey of adult Indian heights with adult American heights shows a similar relationship—Indian men are about 88 percent of the height of contemporary American men, and overall, Indians are among the shortest populations in the world (Deaton, 2007, 2008).

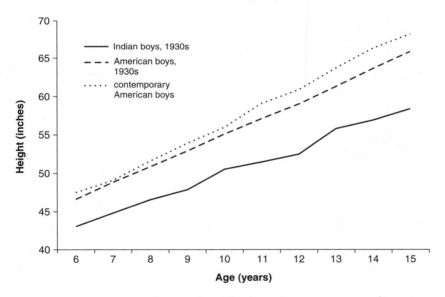

Figure 5.4   Comparison of Indian boys' height in the 1930s compared to American boys from the 1930s and a contemporary sample of American boys. Based on data from Aykroyd, W. R., and Krishnan, B. G. 1936. Diet surveys in South Indian villages. *Indian Journal of Medical Research*, 24, 667–688; and Richey, H. G. 1937. The relation of accelerated, normal and retarded puberty to the height and weight of school children. *Monographs of the Society for Research in Child Development*, 2, i–67.

Aykroyd went on to conduct some small-scale milk intervention studies among mostly adolescent hostel students, the results of which are in
Table 5.1. He noted that milk was generally not provided for students in
the hostels; if it was given it was in very small amounts in tea (Aykroyd
and Rajagopal, 1936a). He found that adding one cup of reconstituted
powdered skim milk to the diet of boys resulted in significantly greater
gains in height and weight compared to a calorically equivalent supplement
of millet. The study of girls was slightly different, as there was no comparable control group, but compared to girls who ate a similar diet, girls who
received a milk supplement grew more in both height and weight. Interestingly, growth in the second three months of supplementation slowed dramatically compared to the first, suggesting that milk offered short-term
gains, but did not continue to spur growth at the same pace that it did in the
beginning weeks of the studies. Thus research in India contributed to the

Table 5.1   Aykroyd and Krishnan's milk supplementation studies of boys and
girls in hostels in South India

	Height change (inches)	Weight change (pounds)
Boys 7–19 (most 11–15 years), N = 122, initial three-month study		
1 ounce of skim milk powder daily = 8 ounces of milk	0.61	4.77
Millet (calories equal to milk) daily	0.35	2.13
Difference	0.26	2.64
Groups were then switched, 10.5-week study		
1 ounce of skim milk powder daily = 8 ounces of milk	0.69	3.07
Millet (caloric equivalent to milk)	0.43	1.10
Difference	0.26	1.97
Girls 10–16 years, N = 35, three-month study		
1.5 ounces of skim milk powder daily = 8 ounces of milk	0.80	4.80
similar diet, no milk	0.53	0.80
Difference	0.27	4.00
Milk group gains after another three months	0.53	1.80
Boys 10–16 years, N = 32, three-month study		
8 ounces of milk (from skim milk powder)	0.67	4.57
with continued supplementation for three additional months	0.59	1.60

Based on data from Aykroyd, W. R., and Krishnan, B. G. 1937. The effect of skimmed
milk, soya bean, and other foods in supplementing typical Indian diets. *Indian Journal of
Medical Research*, 24, 1093–1115.

understanding that milk seemed to enhance growth (at least in the short term, and among adolescent boys) compared to other foods, and, as a result, furthered the cause of enhancing access to milk as a way of improving nutritional status.

Concluding that "the value of milk as food for children is recognized even by the illiterate, but over great areas of the country whole milk is scanty and beyond the reach of the poor," Aykroyd went on to link the lack of access to milk to problems of production: "Unquestionably, it is preferable that Indian children should consume whole milk locally produced. At present, however in many parts of the country, it is a case of imported milk or no milk at all. It remains questionable whether, in many areas, demand for milk can create an adequate local supply. It will be some time before cheap standard skimmed milk products, locally produced, become available in quantity in India" (Aykroyd and Krishnan, 1937, 1105).

Thus, the idea that milk was a particularly good food was apparently entrenched (at least in South India where the research was done), and Aykroyd and his Indian colleagues demonstrated that it could enhance child growth, but in keeping with the pessimism about Indian agriculture that pervaded discussions about productivity at the time, there seemed to be no means to provide children with sufficient safe milk. Aykroyd joined Wright and others in the lament that despite evidence of its benefits to child growth, it would simply be too costly to recommend eight ounces of milk per day for school children. This is quite in contrast to the United States, where the dairy infrastructure was elaborated and the problem was getting people (especially children) to consume more of the abundant supply. Indeed a study of Iowa school children in the 1950s found that boys and girls (ages six to sixteen) were not meeting the quart-per-day requirement, consuming "only" between 2.2 and 3.3 cups per day (Eppright and Swanson, 1955)!

In an attempt to ensure better diets, and as a result more optimal physical and cognitive growth and development among children, India has also experimented with school meal programs, with the first "mid-day meal program" introduced in Madras in 1925 (Chutani, 2012). Free school lunches were offered in Tamil Nadu in the 1960s and expanded in the 1980s. In 1995 the federally funded "National Programme for Nutritional Support to Primary Education (NP-NSPE; commonly known as the "Midday Meal [MDM] Programme") was inaugurated and provided 100 g of food grains per student per day. This scheme was converted into an entitlement for a cooked meal in 2001, and meals generally included a grain, pulses, vegetables, and, when possible, fruit, collectively providing the 450 calories and 12 grams of protein mandated by policy revisions in 2006 (Chutani, 2012).

Compared to the U.S. school meal program, the MDM program is notable for its lack of milk, reflecting the fact that the program in the United States was strongly motivated by a desire to find an outlet for the surplus milk produced domestically. In India there is no such surplus, and the government has been unwilling to commit to buying up or importing milk for its schoolchildren. Critics have noted the lack of milk in both the MDM program and in the Integrated Child Development Scheme Supplemental Nutrition Program (ICDS-SNP), which provides additional food, mostly grains and pulses, for pre-school age children, and includes take-home rations (Working Group on Children under Six, 2007). Pointed comments have been made about the inclusion of micronutrient supplements or fortified rations in SNPs, when milk or eggs are local "natural" foods that could have been subsidized. This promotes "the notion that special 'medicalised' and expensive food is required to deal with micronutrient deficiencies. While there is, on the one hand, a decision not to spend on more expensive 'natural foods' such as milk or eggs, there is no hesitation to spend much more on micronutrient supplements of this kind" (Mishra et al., 2009, 265). Fortified items aside, the program is in line with nutritionists' views that the fundamental need for children is enough calories, and not protein-rich foods in particular (Gopalan et al., 1984).

Milk and milk products are featured in the 2010 edition (and previous versions) of the National Institute of Nutrition Dietary Guidelines for Indians, which recommend 200 to 300g of milk per day for children (National Institute of Nutrition, 2010). Although consuming more milk or dairy products is not spelled out in the key 14 points, point 1 ("Eat a variety of foods to ensure a balanced diet") is elaborated with the statements: "Milk that provides good quality proteins and calcium must be an essential item of the diet, particularly for infants, children and women. Include in the diets, foods of animal origin such as milk, eggs and meat, particularly for pregnant and lactating women and children." For children and adolescents, "plenty of milk and milk products" should be consumed, primarily for their calcium content. Milk and milk products are among the "body building and protective" foods, and as such are particularly recommended for children. Although the guidelines state 200 to 300g of milk per day, the sample meal plans for children and adolescents include an additional glass of milk per day, or 500g in total. Notably, the National Institute of Nutrition report indicates only a 245g per capita per day availability of milk and sadly acknowledges that there would not be enough milk for all Indians to meet this recommendation.

In their comprehensive *Nutritive Value of Indian Foods,* also published by the National Institute of Nutrition (founded by Aykroyd's predecessor, Robert McCarrison) Gopalan and colleagues affirmed that "the best source of animal protein for growing children is milk," but they too lamented that "economic considerations often preclude the inclusion of milk or other animal foods in adequate amounts" (1984, 6). Milk is highlighted as a source of high quality protein and also calcium, and an "ideal food for infants and children," although green leafy vegetables and the millet *ragi* are also singled out for their high calcium concentrations. As Gopalan and colleagues noted,

> these amounts [recommendations] are low, but it should be pointed out that these low figures are suggested as practical levels in the context of the prevailing low per capita availability of milk in the country. It should be our aim in food planning to achieve a much higher figure than this. In the more advanced countries and also in some regions in our country, the daily intake of milk is nearly 600 ml per person [and requirements are for ~750 g]. Renewed and vigorous efforts should be made to increase the average level of milk consumption, and in the meantime the available milk should be channelized to meet the priority needs of infants, growing children and pregnant and nursing women. (1984, 26)

In the context in which average intake was recognized as less than 100 g (< 1/2 cup, a level that remained in 2006, and reported in the 2010 Dietary Guidelines), recommending even a cup of milk (250 ml) was recognized as suboptimal, but "even a little milk is better than none" for children, as "milk is an ideal food for infants and children" (Gopalan et al., 1984, 42). Data indicate much lower intake of milk and milk product among rural children especially, where the youngest (age one to three years) consumed an average of only 72 g of milk per day, an amount that dwindled to ~50 g per day and remained consistent among children up through eighteen years (National Nutrition Monitoring Bureau, 2002). This broad survey found no indications of gender differences in milk or milk product intake among children of any age.

Body Size and Milk: Meanings of Growth

If milk is promoted for its "special" growth-enhancing qualities, it is worth considering the meanings attached to growth in both countries. What does height symbolize? From a public health perspective, height has long been used as a major index of population health (Steckel, 2009; Tanner, 1982). Thus rates of child stunting (having a height-for-age that is two standard deviations lower than the mean of a growth standard), as well as average adult heights, correlate well with standards of living and overall life expectancy across countries (Deaton, 2007). Wealthier—and generally more politically and economically powerful—countries generally have taller citizens. As shown in Figure 5.5, within countries, height is strongly positively correlated with educational attainment and income (Case and Paxson, 2008; Deaton and Arora, 2009). More rapid growth in height during the early childhood period has also been linked to better cognitive performance (Murasko, 2013; Yang et al., 2011), including among children in India (Spears, 2012). Thus the conditions that impact early growth can have long-term consequences for life chances, and parental as well as governmental concerns about child growth are well warranted.

Average height is strikingly different between India and the United States. These differences were described for boys in the earlier part of the twentieth century, and persist in the present. Adult heights in the United States average ~177 cm for men and ~162 cm for women (McDowell et al., 2008), compared to ~155 cm and ~152 cm, for men and women, respectively, in India (Deaton, 2008). In the 2005–2006 round of the Indian National Family Health Survey (INFHS) 48 percent of children

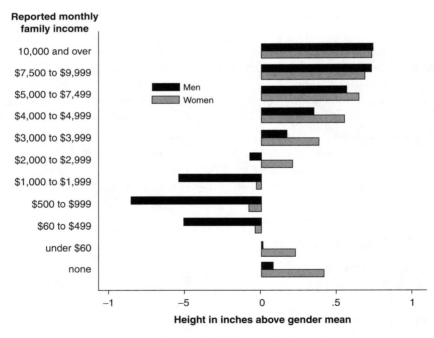

Figure 5.5   Height by income category, United States. Reprinted from *Economics and Human Biology*, 7(2), Angus Deaton and Raksha Arora. Life at the Top: The Benefits of Height, 133–136. Copyright (2009), with permission from Elsevier.

under the age of five years were stunted in height and 40 percent were underweight, although these indices varied by state—only 25 percent of children up to five years were stunted in Kerala, but over 56 percent were so in the most populous state of Uttar Pradesh. Urban children were less likely to be stunted than rural children (40% versus 51%) (Kanjilal et al., 2010). Stunting worsens with increasing age. In the first six months around 20 percent of infants were stunted, but this increased to 58 percent among eighteen- to twenty-three-month-olds. Between two and five years, these level off at just over half (51%) (Gopalan and Ramachandran, 2011). These indices all contrast markedly with those from the United States, where stunting occurs among less than 5 percent of children, although there are differences by poverty status (Lewitt and Kerrebrock, 1997; Markowitz and Cosminsky, 2005), and being overweight is a much more common phenomenon than being underweight among children. Importantly, these differences in height are not the result of fixed genetic differences between Indian and U.S. populations. South Asian children raised in the United States or Europe grow to the same size as local children (Deaton, 2007).

The link between height and political and economic power was not lost on either those in the subcontinent agitating for independence from Britain, nor on American presidents and nutritionists in the first half of the twentieth century. Milk's particular benefits as a food to promote physical growth and strength, as well as its scarcity, were taken up in the nascent Indian Independence movement, and were intertwined with cow protection movements. As Charu Gupta noted, "The cow was now [in the 1920s] more directly linked with building a strong nation, a nation of Hindu men who had grown weak and poor from lack of milk and *ghee*. For a body of healthy sons, cows became essential. . . . Like a mother, she could feed her sons with milk, making them stronger" (2001, 4296). The need for protection and improvement of cows to produce milk would correct "the poor physique of many of the population" (Home Poll 1922, quoted in Gupta, 2001, 4296). Stunted growth of Hindus was attributed to the shortage of cows, created by the British predilection for beef and Muslim sacrifice of cows in celebration of Bakr 'Id (Adcock, 2010).

Thus links between cow milk, national strength, and physical growth were articulated as India moved toward statehood. Shortages of milk were decried, but despite the greater milk productivity of buffalo, the cow proved a more compelling symbol around which the Independence movement could rally (see Figure 4.4). However, after Independence the cow did not persist as a national symbol—instead India's national emblem was taken from the Lion Capital erected by the Emperor Ashoka. In the emblem, three lions sit atop a wheel, on which there is a relief of a bull, elephant, horse, and lion, but no cow is featured. The zebu cow, often undernourished and giving little milk, did not provide a compelling symbol of the hopes and dreams of the new nation, and more powerful animals are used to represent independent India. In her work on animal images used as national symbols, Radhika Parameswaran has noted how the cow, the animal so closely associated with India, has been conspicuous in its absence on international magazine covers that showcase India's rise to global prominence over the past 20 years (Parameswaran, 2011).

As those agitating for Independence were aware, body size, especially height, not only describes the physical size of a country's citizenry, but also its strength and power. These ideas resonate as well in contemporary China, where there is explicit rhetoric supporting the expansion of milk drinking as the means to make up for "growth deficits" among Chinese citizens compared to Europeans or Americans (Wiley, 2007, 2011a). Average heights of Chinese are already greater than those of Indian citizens (166–170 cm for men and 157–159 cm for women; Yang et al., 2005). In both India and China, the growth metaphor is a powerful one: both countries have had rapidly expanding economies and are flexing their growing political muscles. Of course, the growth trope must be carefully crafted as distinct from population growth, which is very much out of favor in both countries. In China, milk provides a means to achieve improvements in "population quality" (*renkou sushi;* Jing, 2000), and Prime Minister Wen Jiabao was quoted as saying "I have a dream to provide every Chinese, especially children, sufficient milk each day" (British Broadcasting Corporation, 2007). Perhaps an indication of differences in their governmental structures and agendas, the Indian government has shown no commitment to expanding access to milk to its children, while the Chinese government has made explicit commitments (at least rhetorically, and in some areas it has been put in place, although fatal cases of contaminated milk have marred this campaign's triumphs).

India's upwardly mobile urban citizens are embracing milk through their own expenditures, as evidenced by growing levels of household purchases and consumption (Ali, 2007). Milk advertisements tend to feature the salutary effects of milk drinking on child health and growth to encourage milk purchases among this growing demographic. A particularly direct ad campaign by the national milk cooperative Mother Dairy features a variety of children dressed in oversize adult professional clothing, with the statement in bold, capital letters: "The Country Needs You! Grow Faster" (see Figure 5.6). Nandini, a milk cooperative in the southern state of Karnataka, uses the Dietary Guidelines for Indians in

Figure 5.6 A Mother Dairy milk ad from India emphasizing milk's contributions to faster growth and development. http://www.motherdairy.com/campaign.asp.

its promotions, noting that "Milk is the perfect beverage for today's kids and teens" (http://www.kmfnandini.coop). Mother Dairy's ads have been noted for their direct call to children themselves as potential consumers (as opposed to their parents) and the lack of the usual references to health or purity (Rai, 2006). Complan, a fortified milk powder sold by Heinz, comes with a height measuring guide on its packaging, and its web FAQs on child growth and nutrition explicitly state: "When planned as a part of growing child's diet, fortified health drinks such as Complan are a good and effective supplements to help your child grow taller as they meet the child's macro and micronutrient requirements" (http://www.heinz.co.in /complan/did_you_know.aspx).

As noted in the previous chapter, milk advertisements in India also frequently use cows (see Figure 4.6), set amidst a verdant meadow of grass. The cows featured are usually Holsteins, the classic black-and-white cow breed that was domesticated in northern Europe, rather than the beloved, distinctive humped zebu cattle that are native to South Asia. In general Holsteins are a much larger and heftier breed than the ubiquitous zebu and are more prolific milk producers. While there might be political reasons for not featuring zebu cows, it may be that the larger size of Holsteins is the key symbol in these ads: bony zebu dairy cows may not be compelling if milk is being singled out for its growth-promoting abilities. This size differential also corresponds with the fact that Holsteins are Western cows. In the wake of new economic policies in India enacted in the 1990s that opened the doors to multinational corporations (many immediately recognizable as European or U.S. brands), milk, which is largely still a domestic product, could use Western motifs as a way of "modernizing" or "globalizing" their image. This has the further effect of decoupling milk from its traditional (and perhaps "old-fashioned") roots, instead linking it to India's full-fledged emergence into the global political economy.[4]

Larger, more productive cows, and larger, wealthier citizens mark a distinction between India and the United States and countries in northern Europe. A large body (in height and weight) is symbolic of the West and social and economic success, even while obesity (and warnings about its dangers) surges in both contexts. But do Indian parents perceive milk as a means for their children to attain larger sizes? My ongoing research in Pune, a large city in the central Indian state of Maharashtra on the relationship between milk consumption and different aspects of child growth provides some insight into this question. Parents (mainly mothers) of 76 children, who had been part of a study on maternal and child diet and growth since birth, were asked about their understandings of milk in relation to child growth when the children were seen recently at ages of five to six years. Over three-quarters of the children reportedly drank milk regularly. Asked about what benefits milk might have for their children, the most common answer was that it enhanced cognitive development (26%). Over 15 percent noted that it enhanced growth in height, and 17 percent referenced strength and strong bones. Almost one-quarter noted that milk was a good meal replacement for when a child was not eating other foods, or that it induced satiety.

Asked explicitly if they thought that milk enhanced growth, 93 percent said "yes," and a similar percentage indicated that milk had equal benefits for boys and girls—one sex did not require more than the other (and those who said "no" were equally split between boys and girls). Among the benefits of milk, 16 percent noted calcium specifically, with somewhat fewer mothers mentioning energy or micronutrients as a category. Very few (3%) referenced protein. Calcium was most often listed in association with "strong bones." Almost one-third said that if a child didn't drink milk, the child would be weak, lack strength, or have weak bones, and almost one-quarter said the child's growth in height would be compromised.

Roughly half of all children drank cow milk, while the other half drank buffalo milk. Small percentages drank either both or goat milk. About 44 percent said that cow milk was "better" for children, only 35 percent reported buffalo milk as better, while 23 percent said goat milk was a more suitable milk for children. Availability was a major factor in this discrepancy. But it was also the case that half of the mothers who reported that milk was good for cognitive development thought that cow milk was better for children, while only 10 percent thought buffalo milk was better. In contrast, among those who noted milk as contributing to physical growth and strength, more thought buffalo milk was better (37%) compared to cow milk (26%). This is consistent with expressed sentiments that cow milk can make you smarter, but buffalo milk makes you bigger, but dumber, much as buffalo are considered to be less intelligent than cows. A substantial (20–25%) of mothers who noted either cognitive or physical growth benefits suggested that goat milk was actually better for children. This appears to be related to perceptions of goat milk as more nutritious due to the fact that goats eat lots of herbs and shrubs, and that it was more easily digested, rather than any potential impact on strength or smartness.

Thus it would appear, based on this small sample of rural and urban parents, that most view milk consumption as important to the growth and development of their children, especially in the domains of height, strength, and intellect. Nutritionist messages seem to have caught on here, with many specifically referencing nutritive qualities of milk (calcium, vitamins, protein) as the source of growth- and health-enhancing effects, although few mothers mentioned protein as milk's primary contribution to their children's diets, despite this being the nutrient often considered limited in Indian diets. Surprisingly there was no mention of more traditional motifs related to Krishna, despite Krishna's stories being parables about maternal nurturance and his fondness for milk and milk products as a child. Krishna's idealized masculine form does not seem to have influenced milk-drinking practices among children, and, consistent with the data from the National Nutrition Monitoring Bureau (2002), there are no consistent gender differences in milk consumption or in ascription of milk's benefits. In a country known for gender discrimination, manifesting in gender differences in child health (Miller, 1987), this was a surprising outcome, given that milk is an expensive commodity.

What do parents expect from greater size and cognitive performance among their children? Greater height is a highly desirable trait for males and females, particularly among the upwardly mobile middle class. A brief survey of marriage ads from Pune, India, revealed that virtually all—whether it was grooms seeking brides or vice versa—referenced the individual's height, and when the desired height of the partner was indicated, it was always "tall," regardless of sex (Wiley, 2011a). Marriage is a means by which families may enhance their social and economic status (for females by marrying up the caste hierarchy; for males by acquiring a large dowry). Height is a measure of the success of individuals and their families, and especially reflects on the quality of a mother's care (R. Parameswaran, personal communication). Given its positive correlation with socioeconomic status within and between countries (Deaton, 2007; Deaton and Arora, 2009) and the way in which it serves as a symbol of distinction between wealthy and poor countries, height is a potent metaphor for and marker of power differentials. Moreover, in India, the linkage between cognitive performance and height is more than twice as strong as it is in the United States, suggesting that better nutrition in childhood is crucial (Spears, 2012). Cognitive performance in turn is linked to greater job opportunities, and upward mobility. That milk might enhance both of these in one glass is a powerful motivator to provide children with milk. But unlike China, where milk production is quite low compared to India, the Indian government has made no effort to act as the provider, instead leaving it up to individual families or nongovernmental institutions to provide milk to children. On the other hand, lack of access to milk due to price or availability provides a justification for the dairy industry's continued expansion, and if the growth, strength, and wealth of its citizens is enhanced as a result (despite the lack of government assistance in this effort), it will be further evidence of India's political and economic ascendancy in the postcolonial period. Thus the linkage between milk, growth, strength and success resonates from the individual, family, and household to the national level.

But what about these associations in the United States? Here too the wealthier are generally taller than the poorer, but milk consumption among children is not differently patterned by socioeconomic status (Sebastian et al., 2010; Wiley, 2005). Beliefs about milk enhancing child growth and building "strong bones" are ubiquitous, but seemingly not sufficient to bolster milk consumption. Advertisements in the late 1900s (see Figure 5.7) were explicit in their claims that milk would "grow giants." In the early 2000s these were somewhat more specific: "Got milk? Get Tall! Want your kids to grow? The calcium in milk helps your bones grow strong. So give them a tall glass" (Milk Processors Education Program. n.d.). Note the conflation of growing "strong bones" and growing "tall"—these are two different aspects of bone biology (density versus length), but they both index strength and size. Calcium is also positioned as the cause of these benefits, and this association is well known in the United States.

The milk-calcium-bone link has become so well known that milk has become associated in the minds of many Americans as a drink for postmenopausal women to reduce the risk of osteoporosis. At the same time, as the risk of osteoporosis has been increasingly understood as an outcome of the peak bone density and mineral content (which is also related

Figure 5.7 U.S. milk advertisement from the 1970s. "Dean's milk grows giants." Outdoor Advertising Association of America (OAAA) Archives, David M. Rubenstein Rare Book & Manuscript Library, Duke University.

to overall bone size and hence height) achieved in childhood (Cooper et al., 2006), most people can articulate the view that children need to drink milk to "build strong bones" (Wiley, unpublished data). As noted in Chapter 3, milk consumption has been sliding since its peak in World War II, but the decline has been especially precipitous among children. As Figure 5.8 shows, since 1977, children's milk consumption dropped by almost 30 percent, and that of adolescents declined by almost half, while adult consumption did not decline significantly (Sebastian et al., 2010). Average intake has declined in part because fewer individuals in all age groups are consuming milk, with the most dramatic drop among adolescents. About 80 percent of two- to eleven-year-olds drink milk, compared to just over 50 percent of adolescents, and under 50 percent of adults (Sebastian et al., 2010).

Why are children not drinking their milk in America? Most, in fact, are, but they are not drinking very much, with average consumption among two- to eleven-year-olds just over 1.25 cups (~320g) per day (still over six times as much as the average rural Indian child). Do American parents not care about gains in height? Average height increased among Americans born up through the mid-twentieth century, but after that there has been a leveling off of height, and the relationship between

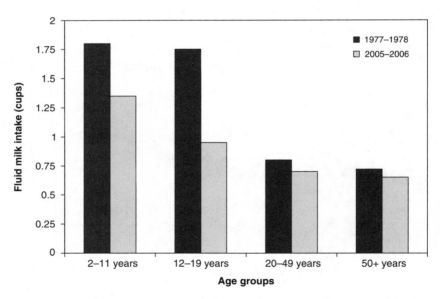

Figure 5.8   Milk consumption trends by age group, nationally representative data from the United States, 1977–1978 and 2005–2006. U.S. Department of Agriculture, http://ars.usda.gov/Services/docs.htm?docid=19476.

height, education, and income has remained relatively stable. Thus Americans are not growing taller, and while they were the tallest national population prior to World War II, there have been greater height gains among northern European countries subsequently, and now the United States lags behind these countries substantially. Some have suggested that this differential is related to differences in health care access and social safety nets across countries (Komlos and Baur, 2004). Others have suggested that declining milk consumption may have contributed to the lack of increases in height in the United States, because milk consumption did not drop as markedly in northern Europe, especially in the Netherlands, where average height increased the most (de Beer, 2012). Nonetheless, Americans are now much taller than they were in the early part of the century, despite milk consumption being the same or even lower (Wiley, 2011c), but the lack of recent gains in height may attenuate any causal link that individuals imagine between milk consumption and growth in height.

What is the state of the science with respect to the relationship between milk consumption and growth in height? A recent meta-analysis suggests that there is a modest positive relationship, which may be stronger when children are undernourished, and among adolescents, who are growing rapidly (de Beer, 2012; see also Hoppe et al., 2006; and Wiley, 2012). Relationships seem to be restricted to fluid milk rather than dairy products in general. There may also be associations with earlier age at menarche (Wiley, 2011b). Thus, while promotions attributing greater height to milk appear to resonate well with, and "make sense" to consumers, the current scientific literature does not provide unambiguous support for these statements.

With widespread understanding of milk as a "good food" for children, and its associations with growing strong bones, these health messages do not seem to bring a urgency to milk purchases or consumption. As a result, advertisements seem to have turned away from a focus on young children and growing strong bones and have hitched themselves to public health initiatives encouraging physical activity. This is ironic on one front: there has long been debate over whether weight-bearing activity or calcium (read: milk) consumption contributes more to bone density (Lanou et al., 2005). On the other hand, since both seem to increase bone density it makes sense to combine efforts, and governmental public health initiatives have done so. The physical activity message is really one designed to confront the child obesity epidemic, and failed attempts by the dairy industry to market its products for weight loss (24 ounces/24 hours "Milk your diet. Lose weight!").[5]

At the same time, however, the demographic featured in milk advertisements is adults, mostly women, as milk positions itself "for all ages."
Here milk is a health problem solver, readily available to ameliorate a
variety of issues, from premenstrual syndrome to colon cancer. With declining fertility and an aging baby-boom population, and with a tired
message (and no immediately tangible evidence) that milk enhances
strong bones or height, milk processors are desperate to find a new market, and the osteoporosis message has exhausted its power to motivate.
However, the ingrained message and intuitive appreciation of milk as a
food that enhances child growth, amidst concerns about American children being "too big" (i.e., too fat) may not be enough to sell more milk.
Thus with concerns about "excess growth," rising rates of obesity, and
chronic disease, the message that "milk makes children grow" no longer
packs the rhetorical punch that it did in the early twentieth century.

The opposing milk consumption trends among children in India and
the United States both represent children's entry into the world of consumers, able to articulate their preferences and influence family purchasing patterns (Curtis et al., 2010). In the United States these preferences
take the form of purchasing other drinks. There are whole supermarket
aisles devoted to alternative drinks—from fruit drinks to soda, a lavish
selection of cleverly and widely advertised, branded, brightly colored,
and pleasingly sweet products—that are clambering for children's attention. Milk must compete with proliferating beverage choices, and with
other drinks fortified with micronutrients, or just promising good taste
and fun, it is difficult to do so. Flavorings, sweeteners, and color have
been added to milk since the early twentieth century to make milk more
attractive and palatable to children, but it can no longer keep up with
the sheer variety and novelty of other (nonperishable) drinks. Yet milk
retains a place in children's diets through school milk programs,where
chocolate milk is the best seller,) and its status as the "original" kid's
food, inherently healthier nutrient profile, and entrenched views about
its benefits to children's bodies keep it in the refrigerator. According to
Charlene Elliott, children realize that the foods they identify as "kid's
foods" are "unhealthy"—but they are "fun," in contrast to adult foods
that are considered healthy but boring (Elliott, 2011). In one sense then,
in the United States milk no longer fits the category of "children's food"
except when flavored, colored, or served with a sweetened, colored cereal. Children's culinary needs have become defined in terms of "fun"
rather than nutrients (Elliott, 2010) in a context in which nutrient deficiencies are rare.

We have little insight into how these forces might be playing out in India. The Mother Dairy ads speaking directly to children suggest that children now might be able to assert more authority over household food purchases. Certainly there is concern that Western foods are replacing traditional foods eaten by children, and that these contribute to the growing problem of child overweight and obesity in India (Sharma et al., 2011). Childhood overweight is estimated at around 13 percent, and obesity at about 3 percent (Midha et al., 2012), with higher prevalence among urban children from wealthier households (Bhardwaj et al., 2008). Nonetheless, milk appears to be part of the mix of Western foods, but whether it will maintain its status among them remains to be seen, or whether its identity as a "traditional" food will be re-established. Sweetened and colored milk is marketed to children and adults alike by such companies as Amul Dairy: "Amul Kool Koko is a fun, nourishing drink with chocolate"; "Amul Kool Milk Shake is milk and sugar blended with fruit or almonds and . . . the product has calcium, protein, carbohydrates, vitamins, etc. for healthy growth of a human body. . . . This product can be consumed by all people irrespective of their age" (see www.amul.com). The "fun" has crept into Indian milk products; although they are not specifically targeted to children, the idea of "healthy growth of a human body" certainly links them to the young. Interestingly, Amul's "plain" milk does not feature this reference to growth on its packaging or in its ads.

In sum, milk as an essential drink for post-weaning-age children became established in the late nineteenth and early twentieth centuries in the United States as efficiencies in milk production and transport generated a glut of milk that needed a market beyond that of infants in need of a breast milk substitute. Selling milk as a food that would enhance child growth at a time when concerns about the unique aspects of child health were coming into popular consciousness "made sense" given milk's evolutionary role to support the growth and development of nursing infants. But as gains in height slowed and new beverages entered the market, milk lost ground as the main drink for children. The dairy industry has tried multiple campaigns in a largely failed attempt to establish milk as an essential part of everyone's diet, not just that of children, who are a shrinking demographic relative to aging adults. Claims that milk consumption enhances height in India have a great deal of traction right now, as Indians are among the shortest populations in the world, a stature incommensurate with their growing "Asian Tiger" economy. In contrast to the United States, the paucity of milk for the market is seen as the culprit behind stunted growth. Taller, larger bodies are a sign of wealth and

health and characteristic of the modern powerful citizenry, but for most
Indians, consumption of valued commodities such as milk must be nego-
tiated among household members. School-age children have not had an
uncontested claim, although that may be changing as children emerge as
active consumers. At the same time, milk is becoming commoditized to
fill a variety of market niches, from "fun" food to nutrient-rich food,
and as such advertisers are banking on a much broader market for their
products.

6

Conclusion
Milk, Biology, and Culture
in India and the United States

What is to be gleaned from this exploration and comparison of the dairy cultures of India and the United States? Clearly the two regions share a dairy culture, where milk and milk products are considered normal and normative parts of the diet and dairy production and its ancillary economic activities make up a sizeable portion of national agricultural economies. This dairy culture ideal stems from different historical sources and is tethered to each country's contemporary social institutions, yielding some important points of contrast. Despite their shared "lactophilia" and normative views of dairy production and consumption, milk drinking falls well short of this ideal, though for quite different reasons, and the two countries are also experiencing divergent trends in milk drinking.

Here I revisit these points of intersection and divergence in the dairy cultures of India and the United States, with a focus on the ways in which milk is perceived as a special kind of food that has been enmeshed in multiple (and often surprising) aspects of social life. This is in contrast to its delimited and specific evolved function in the life history of mammals, although that function (i.e., to support growth) underlies many of its social involvements. Nodes of comparison that provide especially meaningful insights include the genetic basis for milk digestion (lactase persistence), rhetoric about abundance and scarcity of dairy products, as well as wealth and poverty, and how these relate to body size and health, both idealized and real, as well as to differences among milk sources. Milk has played an important role in nation-building projects in both countries, straddles the divide between the "traditional" and the "modern," and it has had the uncanny ability to take on new meanings in light of emerging social trends.

This shared dairy culture exists amidst quite different culinary cultures. The U.S. has long relied on grains (especially wheat), meat, and dairy as dietary staples. In contrast, Indian diets have their foundation in a diversity of grains, pulses, local fruits and vegetables, and dairy. Meat consumption is low, and India has a large vegetarian population: up to 40 percent of the population eschews meat (especially beef) and eggs due to religious, health, or economic reasons (Delgado et al., 2003), but embraces dairy. A recent Gallup poll indicates that only 5 percent of the U.S. population self-identifies as vegetarian, a prevalence that has remained relatively constant over the past decades.[1] As a consequence, although dairy is the main point of intersection in the diets of the two countries, its position and significance is quite different.

One of the key differentiating factors between Indian and American dairy traditions has been the mode in which milk was consumed. Historically it appears that, at least in the northwest, Indians may have been more likely to drink fresh milk than Europeans. When milk was processed further it was through fermentation into yogurt rather than following that with separation of the solids (curds) and liquid (whey) and further aging or brining of the solids into the diversity of cheeses found in Europe (see Figure 6.1). Both traditions feature butter making, but in this case, Indian processing starts with fermented milk, separates the fat from the remaining liquids, and then clarifies the cream to rid the fat of residual solids.

Figure 6.1   Cows, buffalo, ghee, and curds: An idyllic dairy scene from India. © Trustees of the British Museum.

The question has been why there is no aged cheese tradition in India, although this could be interpreted as a Eurocentric bias that privileges this form of dairy consumption. The question has usually been answered by reference to Hindu resistance to "breaking the curds" (i.e., artificially separating the curds from whey), or warm climatic conditions that encourage fermentation but that are detrimental to long-term aging (Mendelson, 2008). Fresh cheese traditions are more common in the north and west, as well as in areas where there was a strong Portuguese influence (Goa, Bengal). From a nutritional perspective, cheese is primarily a source of fat and protein; collectively *ghee* and yogurt provide those same macronutrients, and both *ghee* and cheese preserve easily. When dietary staples are grains or tubers, and meat consumption is uncommon, fats are likely to be limited in the diet. Fats yield more than twice the calories as carbohydrates or protein per gram, and certainly make a bland and starchy diet more palatable. Historically, and in populations with limited resources, fats are highly valued, and the price structure for milk in both the United States and India is based on fat content. Thus both dairy traditions yield fat—an important, and historically limited, component of diets throughout the world (Grigg, 1999)—but have different ways of processing and consuming it.

Genes and Culture: Lactase Persistence in India and the United States

Despite being a "dairy culture," India has overall low rates of lactase persistence, with substantial geographical variation; rates in the United States are higher but variable by ethnic background. Thus substantial numbers of individuals in both countries do not necessarily enjoy drinking substantial quantities of fresh milk. Heterogeneity in lactose digestion is reasonably well understood in the United States. Peoples of northern European descent who predominated in numbers and in political and economic power from the colonial period up through the twentieth century have high rates of adult lactase persistence; Native Americans, slaves from Africa, and later migrants from southern Europe and Asia do not, by and large, share the genes for lactase persistence or an enthusiasm for all things dairy. Each of these non-European populations have culinary traditions that do not include fresh milk usage and, not surprisingly, their milk consumption has tended to remain lower. While the National Dairy Council and other milk promotion groups such as the Milk Processors Education Program (MilkPEP), which sponsored the iconic "got milk?" campaign,[2] have attempted to unite Americans into a common milk-drinking experience by downplaying the role of lactase impersistence as an impediment to milk drinking, milk has not been embraced with enthusiasm. This is not to say that these groups do not see milk as a valuable food, but rather, they follow American norms in both idealizing but at the same time not drinking their milk.

The mutation for lactase persistence that is common in India is the same as that found in high frequencies in European populations, and it appears to have arisen and spread in South Asia independently (Gallego Romero et al., 2012). There is a clear cline with declining frequency from the northwest to the east and south, and this is consistent with the migration patterns of Indo-European speakers with their herds of cattle as they colonized the subcontinent. South India did have a long established indigenous dairy culture that was likely elaborated as Indo-European herders moved south. Yet variation in lactase persistence does not correlate strongly with either linguistic affiliation (i.e., Indo-European languages in the north versus Dravidian languages in the south), or milk consumption patterns at the individual level (Baadkar et al., 2012; Gallego Romero et al., 2012). Broad geographical trends in milk consumption are evident though. The northwest states remain the dairy heartland of India, and both milk production and consumption and lactase persistence rates are accordingly higher there. However, outside of those states, relatively little milk is consumed overall, and these small amounts are not likely to trigger the symptoms of lactose intolerance. But the issue is not indifference to dairy; it is more likely lack of access and frequent utilization of milk in other forms.

Most milk in India is consumed in fermented forms like yogurt or fermented buttermilk, and yogurt is widely taken with meals, especially in the south (Sen, 2004). With a dairy tradition based mainly on cultured dairy products, South Asians would not have had much exposure to unfermented whey, the by-product of cheese making that Europeans appear to have drunk. Whey is rich in lactose and probably provided the main exposure to lactose among pre-nineteenth-century European adults. In India, milk fat was converted into *ghee* after initial fermentation into yogurt; hence the buttermilk left over from *ghee* production would have been fermented and contained lower amounts of lactose. Therefore it is reasonable to imagine that natural selection for lactase persistence was weaker in South Asia compared to northern Europe, and selection would have been stronger among those herding populations that relied most on fresh milk. Lactase persistence frequencies in South Asia are similar to those in the Mediterranean, where yogurt and fresh cheese are more commonly consumed than fresh milk, resulting in relatively low levels of exposure to lactose. However, the very low rates of lactase persistence in east and southeast India and the low rates even in the northwest remain un-explained, and await a more detailed assessment of dairy consumption and historical migration patterns in those regions.

Despite the variability in lactase persistence rates in India, this topic is not widely discussed in nutrition or public health circles, or even addressed by the dairy industry (only the Nandini dairy notes that lactose intolerance is "rare" and hence of little consequence). There are elaborated discussions within the Ayurvedic tradition about milk's "digestibility," which have his-torical connections to similar humoral health belief systems in Europe. Overall, the social, political, and economic context underlying milk pro-motion is very different in India compared to the United States, where the National Dairy Council produces educational materials actively downplaying the significance of lactase impersistence and promoting the "truth about lactose intolerance," while providing ways that individuals who are lactose intolerant can "tolerate" more lactose in their diet or consume dairy products with minimal lactose content.[3] All of these ef-forts are in the service of increasing consumer purchases of dairy prod-ucts, especially milk. The institution roughly analogous to the NDC in India is the National Dairy Development Board (NDDB), whose 2011 annual report does indicate interest in the development of lactose-reduced milk to be marketed to those who are lactose intolerant. But the NDDB is primarily concerned with boosting production rather than cre-ating a market, as demand for dairy products is currently growing at 8 percent per year—twice the rate of production (Singh, 2011). The main critical issue for the Indian dairy industry is accommodating consumer desires for more milk. Hence they exhibit little worry about creating a market or addressing concerns about lactose intolerance that are not broadly articulated.

According to Indian dairy industry analysts, "The major factors driv-ing growth in milk consumption are increased demand due to population growth, growing household incomes, increased demand for value-added milk products, and the preference for fluid milk as a principal protein source across all age groups in India" (Singh, 2011). But if demand is considered to be primarily for fluid milk—as opposed to other forms of dairy—discussion of lactose intolerance may start to move to the fore as individuals who formerly consumed fermented dairy products find them-selves having trouble digesting large amounts of fresh milk. Still, with low average incomes and rapidly rising prices for milk, milk is poised to remain a middle-class beverage, consumed more often by children (who may still be producing lactase), and out of reach of most of the country's population. And to be fair, India faces much more severe problems with access to adequate food by the poor, so concerns about lactose intolerance should not be at the forefront of public health and nutrition. Socioeconomic status is a more significant contributor to variation in milk consumption than lactase persistence status.

Abundance and Scarcity, Wealth and Poverty

The heart of variation between the dairy cultures of the United States and India is the relative abundance and scarcity of milk. As a result, the focus of the dairy industry in the United States is to boost consumption of the copious milk supply. In India it is to boost production in order to meet consumer wishes. In the early colonial period, the United States was already established as a land of abundance. It was not lost on the early colonists that their cows thrived in the New World, many even becoming feral as their populations burgeoned (Crosby, 1986). Producing enough milk was seemingly never in question, and easily accommodated consumption needs, which were primarily for cheese and butter and fresh milk for cooking. In the nineteenth century the issue became one of how to get the ready supply of rural milk to growing cities to meet the demand for fresh milk by those who could no longer produce their own. While peri-urban dairies seemed the solution to that problem, their insalubrious conditions posed serious health risks to this new mode of dairy consumption and they quickly became the target of social reformers seeking to address urban poverty and the high rates of infant and child mortality associated with the consumption of this "swill milk." Urban milksheds thus expanded their radius well beyond city limits, and innovations in refrigerated transportation in concert with pasteurization mandates reduced milk's contributions to infectious disease mortality. Simultaneously this fostered efficiencies in production that readily met demand and then some. New surfeits of milk necessitated the establishment of the NDC to expand the market for fresh milk. In the post–World War II period, government commodity programs were instituted to buy up milk to provide price supports, and domestic programs were put in place to solidify consumption outlets for milk and dairy products, including school feeding programs. Some of this surplus was converted to dried milk and sent overseas as part of food assistance programs established by the Food for Peace Act (also known as PL 480), which provided India with large amounts of food aid in the 1960s.

Similar policies and milk surpluses occurred in northern European countries, and formed the basis for India's famed Operation Flood, designed to provide an outlet for surpluses from Europe and the United

States and stimulate growth in India's dairy industry. Through this aid, the two major dairy cultures of the world were linked in ways that they had not been during the Raj's presence in India, and it seemed that they would both be awash in a wealth of milk. It is ironic that as a result, India is now the largest milk producer in the world, but production is still not adequate to provide sufficient milk to meet dietary recommendations or consumer demand. Rhetoric about food production deficits has dogged India since the pre-Independence days and, despite dramatic economic growth and gains in average income, roughly one-third of the population is considered impoverished and India accounts for 25 percent of the world's hungry population.[4] Government commodity programs exist to maintain prices sufficiently attractive to farmers, but these have long focused on grains to the exclusion of other protein-rich foods such as pulses or dairy products. There are no special children's food programs that provide for milk at the national level. Thus while India may indeed be a major dairy producer and support milk culturally, there is not currently, nor has there been since the Operation Flood days, the political will to make milk a cornerstone of consumption and have India "join the ranks" of the milk-drinking nations. Nor have there been concerted efforts to make such "sacred" foods as milk available to the poor. India is now tentatively entering the dairy export market with the lifting of export restrictions in 2012, which will further reduce domestic availability.

Access to safe and adequate milk has been a rallying cry for antipoverty movements in both India and the United States. With milk positioned as essential to the diet, lack of access to it came to symbolize the essence of poverty, powerlessness, and want. Of course dairy products were foods for the lower classes in Europe—the "white meats" consumed by the poor compared to the "red" meats eaten by those of greater means. In contrast, milk and dairy products are the food of the well-off in India, who are part of the "livestock revolution" characterized by increased incomes, spending power, and greater consumption of animal-source foods, including fluid milk (Delgado, 2003).

Milk and the Body

This contrast of excess milk and its scarcity speaks to a broader discourse about consumption patterns in wealthy versus poor countries. In the former, food is cheap and dense in calories and protein so that overconsumption is a problem, while in the latter, households spend a much greater proportion of their income on food, and getting sufficient nutrients as well as calories from this food is far from guaranteed. In turn, these consumption patterns have been associated with quite different biological outcomes: obesity and its attendant health complications in the United States, and undernutrition, stunted growth, and its associated limitations in India. This is not to say that over- or underconsumption of milk and its products is the cause of these, but claims to that effect have been made. For example, in the United States in the 1970s, when saturated fat came under scrutiny as a contributor to cardiovascular disease, a major campaign ensued to encourage Americans to consumer lower-fat milk and dairy products, and Americans by and large complied, at least with respect to milk, although their high-fat cheese consumption increased in tandem, such that the net effect on dairy intake was nil (see Figure 2.3). Current dietary guidelines encourage consumption of low-fat milk and dairy products, although only low-fat milk and yogurt are widely consumed. Yet evidence for negative effects of high-fat dairy intake on chronic disease or mortality are not compelling, and there is some counterevidence that dairy intake overall is associated with lower mortality from cardiovascular diseases (Elwood et al., 2010).

In India, the "poor physique" of Indians in the early twentieth century was attributed to a lack of access to milk, and milk's importance as a protein source in a population with a relatively high percentage of vegetarians has been highlighted. At the same time, the rising numbers of middle-class Indians who can easily afford milk are also experiencing a dramatic rise in rates of obesity and diabetes (Shetty, 2012), linking their biological outcomes with those of citizens of the United States, although milk consumption is not likely to be the primary mediator of these conditions.

Milk is currently deployed to address the opposing concerns of over- and undernutrition, but this reflects a century-long history of milk being considered—perhaps more so than any other food—a "problem-solving" food, one whose nutrient richness can ameliorate any health concern, from rickets in the early twentieth-century United States (through fortification with vitamin D), growth stunting from undernutrition (through many nutrients, including vitamins, protein, and minerals), osteoporosis (though calcium), and more recently, various cancers, overweight, and diabetes, although the mechanisms for the latter health concerns are not established. In India milk is portrayed as a solution to the problems of growth deficits through claims about its ability to enhance growth in height and cognitive development. Both attributes are seen as essential in the modern economy and the more traditional marriage market, where tallness and educational attainment are desirable attributes of spouses of both sexes, and where a successful marriage brings social and/or economic benefits.

Milk as Modernity and Tradition

The promise of enhancements in physical and cognitive growth and development may ultimately be behind the contention, in a 2007 *New York Times* article, that milk is the "mark of new money" in developing countries such as India (Arnold, 2007). The goal of achieving larger body sizes may underlie a dietary trend in developing countries marked by the "livestock revolution" of increased consumption of animal-source products (Delgado and Narrod, 2002). In contemporary India and in the late nineteenth century in United States, when milk intake also surged, stunting from undernutrition and infectious disease are and were common, and hence evidence of enhanced physical growth provides confirmation of economic and social success. In those contexts, fresh milk—consumed in a new, modern way (produced and marketed by science-based and technology-driven pasteurization, refrigeration, and knowledge in nutrition)—became the means to new ends. The fact that politically and economically powerful nations currently tend to have the highest levels of milk consumption and the tallest citizens creates a package of meanings for milk as a "modern" food, one essential to success in the twenty-first century (Wiley, 2011a). Milk appears as a solution to the problem of past growth deficits and poverty by creating a tall and strong citizenry, and milk's biological role—to sustain the rapid growth of infant mammals—makes such a connection seem intuitive. Importantly, such attributes accrue only to milk as a fresh fluid and do not seem to extend to dairy products more generally. And as the social reformers of the mid-nineteenth century noted, access to clean, pure milk would go a long way to mitigating the negative effects of urban poverty. Despite these contemporary appeals to the Indian middle class, the "problem" that milk was to solve in India established during Operation Flood was rural poverty, but similar to the United States, the primary destination for milk procured from rural areas was the cities, which were also to be rid of their livestock.

Meanwhile, in the United States, where increases in height have slowed, the "milk makes children grow" message no longer has much traction. With a still growing, and young population, this message and the focus on children makes sense in India; with an aging population in the United States, it is not surprising that the focus has shifted to adults with their attendant "diseases of modernization." This marks an attempt to reinvigorate the milk message in relation to contemporary public health concerns: chronic diseases manifesting in later adulthood. While the marketing campaign 24/24 (24 ounces of milk each 24 hours) for weight loss was pulled by the Federal Trade Commission in 2007 for lack of evidence, it marked milk's entry into public health discourse about chronic disease, one that was made possible by the wider consumption of fat-reduced milks, as noted above. However, after the abrupt departure of overt claims about milk's contributions to weight loss, ads emphasizing milk's associations with decreased risk of chronic disease have dwindled in the major milk promotion venues (websites, magazine ads). Certainly concern about chronic disease has not diminished, and the demographic at risk is enlarging as baby-boomers become senior citizens, but, in my view, milk's biological attributes as a food to support the growth of the young prevent it from being fully embraced for whatever benefits it may have in relation to chronic disease (Wiley, 2011a). In other words, milk's biological "specialness" binds it to the young, who can be shaped through milk consumption to be model citizens of the twenty-first century.

In both the United States and India, milk has the status of both an "age-old" and modern, technologically sophisticated food, engineered to meet the nutrient needs of the decade and the demographic with the most potential for growth. Abundant fresh milk, consumed cold from refrigeration and with more guarantees for safety, is a middle-class novelty in India; in the United States this pattern is long established and milk must be re-engineered (or at least remarketed) as a new value-added product: "'Consumers are indicating that they value . . . heart health and better-for-you attributes—and appear willing to pay higher premiums for nontraditional products,' notes MilkPEP CEO Vivien Godfrey. 'We're seeing exciting innovations that add ingredients like antioxidants, fiber, omega-3 fatty acids and whey proteins to dairy products'" (quoted in Goldschmidt, 2012). At the same time there is renewed interest in locally produced and/or organic milk sold in the traditional recyclable glass bottles (Roth et al., 2008). Thus milk is being positioned to solve the problems of modernity through fortification and fat removal, while simultaneously being recast as a local and traditional food, harkening back to an idyllic, healthier past.

Miracle Milk

Milk's specialness can easily morph into "miraculous." As "nature's most nearly perfect food" (Crumbine and Tobey, 1929), cow milk was written up in a 1955 issue of the *Journal of Dairy Science* as the elixir of life (Rusoff, 1955). "Within the last quarter century science has been peering into a drop of milk and has discovered the 'miracle'—for no other food in the world can compare with milk in its outstanding nutritive values" (Rusoff, 1955, 1057). Milk's miraculous properties have been couched in nutritional terms since the early twentieth-century discovery of micronutrients and their importance to human health, and these have been assumed to underlie the political, economic, and physical power and size of dairying civilizations (Crumbine and Tobey, 1929; McCollum, 1922), although India's status a struggling newly independent country in the 1950s, but also one with a long-standing and elaborated dairying tradition, was conveniently overlooked. A very different kind of miracle involving milk occurred in India in September 1995. Statues of Hindu gods were reportedly "drinking" the milk offered to them by worshippers, with the first reported in New Delhi at a shrine to Ganesh. The event lasted less than a day, but news of it spread rapidly across the Hindu world and milk sales in India skyrocketed. Discussion of the incident was framed in terms of religious beliefs versus a scientific explanation of uptake by capillary action in porous material. Although this incident was not about milk per se,[5] it would seem to confirm milk's divine associations and similar renown as the "elixir of life" (Ghatwai, 2008).

In addition to these seemingly elusive aspects of milk's "specialness", both in India and the United States milk has come to be understood and discussed in nutritionist terms, with its nutrients front and center in public health and popular discourse. Unique historical understandings of milk and its role in the diet are disappearing as part of the global spread of nutritionist messages (Wiley, 2011a) and the Western diet in general (Popkin et al., 2012). Nutritional benefits of milk underlie its privileged position as an essential food in dietary guidelines in the United States and India. In the former, however, dairy industry interests are actively articulated in the Dietary Guidelines for Americans due to their oversight by the U.S. Department of Agriculture, which also prioritizes American agricultural interests (Nestle, 2002). Links between the National Institute of Nutrition's dietary recommendations and the NDDB in India are more opaque, but then again, problems are framed in terms of the need to establish an adequate and affordable supply rather than the need to persuade the public of milk's benefits. The lack of milk in government food assistance programs provides further evidence of a more disinterested relationship between the dairy industry and dietary guidance there—at least in terms of encouraging consumption across the whole spectrum of the Indian population.

The Sacred Cow

Much has been made of the "sacred cow complex" in India, but such anthropological scrutiny has not been turned on the cow in the United States, especially in relation to milk consumption. To use the term "sacred cow" metaphorically, it is milk that has this status in dietary guidelines and in American refrigerators, and cows are the beloved source of this goodness. Happy dairy cows (rarely beef cows) are a standard feature of children's books. These cows have many magical features: they talk, fly, type, dance, sing, ride bicycles, and experience the full range of human emotions. Although whimsical, these books emphasize cows as the source of milk, reinforce ideals and norms of milk as an essential food for children, and reify a wholesome, happy rural past. Pictures of cows living in the homes of their human owners offer another view of the dairy cow as something cherished above and beyond their milk. Seen in this light, the cow's sacred status among Hindus may seem less "bizarre" to Americans. Indeed cows are portrayed in Indian children's stories mainly as economic entities, providing an expensive and valued commodity. If anything, they are less likely to have magical qualities. With most of India involved in the hard work of subsistence agriculture, such books are less likely to romanticize rural life. To some extent the tales of Krishna's life do present an idealized rural existence (he is, after all, a god), but the dairy cows form the backdrop for Krishna's lighthearted antics, and his later more serious actions as the defender of *dharma*. In these stories, cows are neither animated creatures with magical powers, nor is their milk associated with such benefits.

What remains mysterious is the buffalo's less exalted status, given their economic superiority to cows, at least with respect to producing milk, and milk with double the fat content. This suggests that the main underlying value of cows is that of bullocks for plowing and other agricultural tasks, rather than as a source of milk. It also demonstrates the force of ideology as a determinant of value. But despite claims to the contrary (Velten, 2010), the dairy culture of South Asia is probably not closely related to Hindu belief systems about the divine status of cows, and cows as a source of milk do not form the basis for materialist interpretations of the sacred cow complex (Harris, 1966). At the same time, cow milk is clearly more highly valued in relation to the Ayurvedic humoral conceptualization of the body, as a more easily digestible form of milk, especially for children. My work with mothers and children in Maharashtra suggests that there is a modest bias toward cow milk as the ideal milk for children, but this did not translate into more children drinking cow milk than buffalo milk. In any event, while buffalo produce more milk and their milk has more fat, which increases its price in the market, biases against them are evident in their absence from milk advertisements, and in folk wisdom about their personality traits.

Cow milk has also taken on a political role, insofar as cow protection served as a form of resistance to the Raj and as it currently serves those with a Hindu nationalist political agenda (Ghatwai, 2008). Although the pre-Independence rhetoric emphasized the lack of cow milk as contributing to physical deficits, current discourse does not frame cow milk's distinct benefits in terms of growth, but rather in relation to Ayurvedic principles and the importance of defending the cow's sacred status. In the latter context buffalo milk is denigrated, and folk wisdom also indicates a bias against buffalo as "dull-witted," a characteristic that the consumer might take on. Thus drinking cow milk is also a path toward cognitive development and full participation in India's economic boon, although the milk available from an urbanite's local store is more likely to be a blend of buffalo and cow.

From the U.S. perspective, differentiating between bovine sources of milk is not necessary, but there are marked parallels in the ways that cows are privileged in the political, economic, and consumer spheres. Since killing a cow or eating beef is most certainly not a crime in the United States, "cow protection" movements have the goal of protecting the politically powerful beef and dairy industry interests. Furthermore, if a consumer is going to consume "real milk," it will almost certainly be from cows, and cow milk consumption has long been used to celebrate white racial superiority and nationalist achievements (Crumbine and Tobey, 1929; McCollum, 1922; Patton, 2004; Rusoff, 1955). But the dairy industry's commitment to describing milk's virtues in nutritionist terms has paved the way for other fortified "milks" packaged and marketed like "real" milk to enter the dairy case. Calcium-, vitamin D–, and vitamin A–fortified soy, almond, coconut, and other milks are now easily available in U.S. groceries. A contending "cow protection" movement has emerged from animal rights groups such as PETA, although their political influence is negligible.

Anxieties about Milk

Milk may be a sacred substance or one with properties unmatched by other foods, but its valuation has always been tempered by fears about its quality (Levenstein, 2012). Milk can spoil rapidly, and animals that produce it can harbor diseases to which humans are vulnerable (tuberculosis, brucellosis).[6] It can be diluted or adulterated, sometimes with fatal consequences (as in the 2008 melamine adulteration of baby formula in China). It can be too expensive or scarce in seasons when cows are not in milk. It can be high in saturated fat, which was identified—rightly or wrongly—as a contributor to cardiovascular disease and obesity in the 1970s (Taubes, 2001). Milk is routinely assumed to be diluted to maximize producer profits in India, and a 2012 survey showed that most milk sampled from around the country did not meet food safety standards. In seven states all of the milk samples showed evidence of contamination. Some was merely diluted while other samples showed evidence of glucose, skimmed-milk powder, detergent, hydrogen peroxide, or urea (Singh, 2012). Consumers have come to expect that they will be cheated by their milk vendors, and haggling over the cost or quality of milk is a common cultural phenomenon among urban housewives. Concerns about the use of rBST in the United States to stimulate milk production in dairy cattle has raised fears among some consumers that this might increase the risk of health problems such as cancer due to the presence of higher amounts of insulin-like growth factor I (IGF-I) in such milk.[7] Such anxieties take on grave import when the food in question is deemed essential, especially in the diets of children, and has official governmental imprimatur. When milk becomes the means by which individual, household, or national goals can be achieved, it is not surprising that reminders of its vulnerability provoke consumer disquiet, but ideas about milk's necessity in the diet, and its close connection to growth and development, drive them to risk consumption anyway or look for alternatives that promise the same (or better) outcomes.

But milk has long been "doctored" to encourage greater consumption, or to enhance its nutrients (e.g., adding vitamin D or vitamin A to reduced-fat milk). Most important, it has been sweetened for children through the additions of chocolate, vanilla, strawberry, and other fruit flavors, and even the addition of ice cream or soda. One cup of milk has about 13 grams of sugar; commercial chocolate milk generally has almost twice that, which is roughly equal to the sugar in a similar amount of any regular soda. Milk drink brands Horlicks, Bournvita, and Complan in India provide similar amounts of added sugar, and some mothers in my study in Maharastra noted that this was the only way in which they could get their children to drink milk. With the rise of intake of sugar-sweetened beverages, and studies targeting them as contributors to high rates of obesity and diabetes (Malik et al., 2010; Popkin and Nielsen, 2003), milk's privileged status as a nutrient-rich alternative to sodas or juices is coming under greater scrutiny, with some school districts banning chocolate milk in school meal programs (Hoag, 2011). In both India and the United States, bitterness is another taste paired with milk, via tea, coffee, or chocolate, often with added sugar. In both cases, milk takes on almost drug-like qualities, and consumers are urged to use whatever means necessary to drink it down.

A Nation-Building Food

Because of milk's "special" qualities, which seem to attach primarily to its fluid state, it has been deployed in India and the United States in a number of "building" projects: individual bodies, families, dairy industries, nation-states. The particular meaning of milk in relation to these projects has varied across these two large dairy cultures. Milk's means to build a strong independent nation in postcolonial South Asia utilized the trope of the Hindu sacred cow. It was cow protection movements mounted in resistance to the British (and Muslim) predilection for beef, which were also meant—at least rhetorically—as a way to ensure enough milk to grow the emergent nation's citizens, although zebu cows were known to be poor producers compared to the non-valorized buffalo. Not enough milk was the lament, whereas in the emerging industrial United States, the surfeit of milk required a new outlet and a new form of consumption: fresh fluid milk as a food for all, not just for infants in need of a breast-milk substitute. In both contexts, poor growth due to food scarcity or lack of affordability combined with infectious diseases of childhood made milk's growth-enhancing qualities especially attractive, although adulteration or contamination remained a concern. Large and powerful citizens were needed to build strong nations—milk consumption reached its apex during World War II in the United States and was marketed at the time as a way to maintain the strength of soldiers and civilians at home (Wiley, 2011c). Fresh milk consumption is rising among the Indian middle class to build citizens with larger sizes and resumes in order to match the growth of Indian political and economic power in the global realm.

Official policy endorsing milk consumption exists in both places at the national level through sanctioned dietary guidelines, but these exert a stronger force in the United States where they are designed in part to maintain a market for this commodity. School feeding and supplemental nutrition programs for women and children provide an outlet for milk, but there are no similar nationally funded mandates in India. Although it is the largest producer and consumer of milk, there has been no attempt to build a national identity around milk in India. Instead, in 2012 Indian newspapers asserted that tea would be named the national drink by April 2013, but in the end the commerce department declined to do so, in order to avoid conflict between the coffee and tea industries. Coffee has long been a drink in southern India, while tea is produced and more widely consumed in the north. Although tea is consumed in much larger quantities on a per capita basis, coffee consumption is growing more rapidly, and to give tea status as the national drink would be to fan the flames of long-standing north-south tensions. Amul, India's largest dairy, which markets itself as the "taste of India" did plead for milk's status as the national drink: "Milk is the world's original energy drink for all age groups and for all healthy nations. . . . Everybody knows milk is the national drink" (*The Times of India*, 23 April 2012). While the veracity of the first statement is questionable, milk consumption could be promoted as the means to a healthier India, one that would align India more closely with wealthier (and healthier) nations (although it is worth noting that the country with the highest life expectancy is Japan, which has only modestly higher milk consumption than India, and less than one-third of that of the United States).

In conclusion, it is tempting to see India's current position in relation to fresh milk consumption as reflective of that of the United States at the turn of the twentieth century. In some ways it is, with infectious disease morbidity and protein-energy malnutrition being major threats to child growth, and rapidly expanding urban, industrial, and consumer-based economies. Yet milk is coming to market with a host of other branded beverages, and milk's status as a local, traditional product amidst the multinational, heavily advertised beverages may prevent its consumption from escalating. And, public health nutrition concerns in India are no longer solely about ensuring sufficient food, but are also about overconsumption, including fat-rich dairy products (Shetty, 2012). On the other hand, with the rise in consumption of animal-source foods in general (Delgado, 2003), milk is not just another beverage, but part of a package of foods viewed as protein rich, and height enhancing. But at the present, milk is promoted to enhance growth and development in India—to solidify India's position among the world's "growing" superpowers, and as a "developing" nation, but this promise exists—along with the myriad other contradictions of modern India—while the milk production infrastructure is described as woefully inadequate and "unmodernized." India's milk industry is much more forward looking, with dairy's "promise" on the horizon; America's remains nostalgic. Milk is promoted as an "age-old" remedy for the health problems that stem, at least in part, from chronic overconsumption in the United States (Wiley, 2011a), and milk harkens back to a halcyon "premodern" era. Despite differences in contemporary meanings of milk, both countries share a long-standing dairy culture and bodies shaped by experiences with milk, onto which the hopes and fears of these nations have been projected.

NOTES

REFERENCES

INDEX

Notes

1. Introduction

1. Not all mammalian milks contain lactose. Many of the marine mammals produce milk that has none or very little. Instead, their milks are very high in fat, with some having as much as 50 percent fat, a feature probably related to their cold marine environments and an infant feeding ecology characterized by infrequent nursing bouts.

2. Every individual has two copies of each gene, one from the mother and one from the father. Different versions of a gene are called alleles. One allele can be dominant to the other, meaning it masks the expression of the other. In the case of lactase, the allele that causes lactase to remain on throughout life masks the expression of the one that turns it off; hence only one copy of the dominant allele will lead to lifelong lactase activity.

2. A Brief Social History of Milk Consumption in the United States

1. Currently two forms of pasteurization are commonly used. One is "high-temperature short-time" (HTST). Milk is passed through warming pipes to raise the temperature to 161°F, where it is held for 15 seconds, then rapidly cooled. Ultra-high temperature (UHT) pasteurization raises the milk to a much higher temperature, but holds it there for only one to two seconds. It is often packaged in aseptic cartons, and has a long shelf life without refrigeration. UHT milk has not gained much of a foothold in the United States, although it is widely available in other countries, including in Europe.

2. The problem of the potential interchangeability between high-fat and low-fat dairy foods has not gone unnoticed. The Harvard Healthy Eating Plate constitutes an alternative set of dietary guidelines crafted by the Nutrition Source at the Harvard University School of Public Health. The Healthy Eating Plate has water in a glass next to the plate, not milk. Milk and dairy are limited to one to two servings per day. See http://www.hsph.harvard.edu/nutritionsource/pyramid/.

3. The organization People for the Ethical Treatment of Animals (PETA) filed a lawsuit in 2011 alleging that the CMAB was engaging in false advertising with

their "Happy Cows" ads, and that cows were often mistreated, abused, or suffered from poor health because of the conditions in which they were kept. The
suit was thrown out in 2012.

4. Diversity in Dairy

1. Dean Foods actually spun off its soy milk division (WhiteWave Foods Co.) in
   an initial public offering of WhiteWave stock in late 2012. It retains a majority
   (88%) interest in the company, however (*Wall Street Journal,* 26 October 2012,
   available: http://online.wsj.com/article/SB1000142405297020459850457808054
   0434262504.html).
2. The text of the letter to the FDA can be found at: http://www.nmpf.org/files/file
   /NMPF%20Misbranding%20Letter%20to%20FDA%204-28-10.pdf.

5. Milk as a Children's Food

1. The reverse was also in play: adding milk to chocolate not only made it cream-
   ier, it also afforded chocolate a new claim as a healthy food, one particularly
   well-suited for children (see Chiapparino, 1995).
2. "Popeye," the U.S. cartoon that rose to popularity in the 1930s, featured the
   eponymous sailor as being exceptionally strong due to his high consumption of
   spinach. This is particularly notable, given that milk was being widely marketed
   as having growth-enhancing abilities during that time in U.S. history.
3. See Pollock (2006) for a critique of these historical milk supplementation studies. A more recent, but similarly structured, supplementation study done over
   two years among Kenyan children revealed that milk did not contribute to
   greater height than a simple energy supplement, and that differences in growth
   among supplementation groups were restricted to children who were growth
   stunted when they entered the study (see Grillenberger et al., 2003).
4. Early advertisements and milk labels in the United States also tended to feature large, placid cows grazing in verdant pastures, before images of plump
   and equally placid children began to be featured after the Civil War. Then the
   link between milk and the strong (white), large body became more direct, with
   cows receding into the background (see DuPuis, 2002).
5. This claim was pulled by the Federal Trade Commission in 2007 due to lack of
   scientific support, but it can still be found at: http://deandairy.com/newscenter
   .php?health_nutrition (embedded links to 2424milk.com are now obsolete).

6. Conclusion

1. See http://www.gallup.com/poll/156215/consider-themselves-vegetarians.aspx.
2. See www.gotmilk.com; milkdelivers.org; gotchocolatemilk.com.
3. The National Dairy Council produces education tips for health professionals
   related to lactose intolerance and dairy. See http://www.nationaldairycouncil
   .org/EducationMaterials/HealthProfessionalsEducationKits/Pages/Lactose
   IntoleranceAndDairy.aspx.

4. See http://www.foodsecurityportal.org/india/resources.
5. See http://www.nytimes.com/1995/10/10/world/india-s-guru-busters-debunk-all
   -that-s-mystical.html?pagewanted=all&src=pm.
6. Ironically, cows probably initially contracted bovine tuberculosis from humans
   when they were domesticated (see Huard et al., 2006).
7. The official determination of the Food and Drug Administration is that milk
   from rBST cows is no different, and hence does not need to be differently regu-
   lated or labeled, than milk from untreated cows. While rBST-treated cows may
   produce milk with higher levels of IGF-I, the FDA maintains that these are within
   the range of normal variation and lower than local levels produced in many human tissues. For the full FDA report on milk from rBST cows, see http://www
   .fda.gov/AnimalVeterinary/SafetyHealth/ProductSafetyInformation/ucm130321
   .htm. Less than 20 percent of milk sold in the United States is from rBST-treated
   cows due to consumer concerns that have not been assuaged by the FDA's
   assessment.

References

Abbott, E. 2008. *Sugar: A Bittersweet History.* New York: Penguin.

Achaya, K. T. 1994a. *The Food Industries of British India.* New York: Oxford University Press.

———. 1994b. *Indian Food: A Historical Companion.* New York: Oxford University Press.

———. 1998. *A Historical Dictionary of Indian Food.* New Delhi: Oxford University Press.

Adamson, M. W. 2004. *Food in Medieval Times.* Westport, CT: Greenwood Press.

Adcock, C. S. 2010. Sacred cows and secular history: Cow protection debates in colonial North India. *Comparative Studies of South Asia, Africa and the Middle East,* 30, 297–311.

Ali, J. 2007. Structural changes in food consumption and nutritional intake from livestock products in India. *South Asia Research,* 27, 137–151.

Alvares, C. (ed.). 1985. *Another Revolution Fails: An Investigation into How and Why India's Operation Flood Project, Touted as the World's Largest Dairy Development Programme, Funded by the EEC, Went Off the Rails.* Delhi: Ajanta Publications.

American Academy of Pediatrics Committee on Nutrition. 1992. The use of whole cow's milk in infancy. *Pediatrics,* 89, 1105–1109.

Anderson, E. N. 1987. *The Food of China.* New Haven, CT: Yale University Press.

Ariès, P. 1962. *Centuries of Childhood.* London: Knopf.

Arnold, D. 2000. *Science, Technology and Medicine in Colonial India.* New York: Cambridge University Press.

Arnold, W. 2007. A thirst for milk bred by new wealth sends prices soaring. *New York Times,* 4 September.

Atkins, P. 2010. *Liquid Materialities: A History of Milk, Science and the Law.* Burlington, VT: Ashgate.

Atkins, P. J. 1992. White poison? The social consequences of milk consumption, 1850–1930. *Social History of Medicine,* 5, 207–227.

Aykroyd, W. R., and Krishnan, B. G. 1936. Diet surveys in South Indian villages. *Indian Journal of Medical Research,* 24, 667–688.

————. 1937. The effect of skimmed milk, soya bean, and other foods in supplementing typical Indian diets. *Indian Journal of Medical Research*, 24, 1093–1115.

Aykroyd, W. R., and Rajagopal, K. 1936a. The state of nutrition of school children in South India. *Indian Journal of Medical Research*, 24, 419–437.

————. 1936b. The state of nutrition of school children in South India, Part II: Diet and deficiency disease in residential hostels. *Indian Journal of Medical Research*, 24, 707–725.

Baadkar, S. V., Mukherjee, M. S., and Lele, S. S. 2012. A study on genetic test of lactase persistence in relation to milk consumption in regional groups of India. *Genetic Testing and Molecular Biomarkers*, 16, 1413–1418.

Babu, J., Kumar, S., Babu, P., Prasad, J. H., and Ghoshal, U. C. 2010. Frequency of lactose malabsorption among healthy southern and northern Indian populations by genetic analysis and lactose hydrogen breath and tolerance tests. *American Journal of Clinical Nutrition*, 91, 140–146.

Banerji, C. 2006. *Feeding the Gods: Memories of Food and Culture in Bengal.* London: Seagull Books.

Basu, A., Mukherjee, N., Roy, S., Sengupta, S., Banerjee, S., Chakraborty, M., Dey, B., Roy, M., Roy, B., Bhattacharyya, N. P., Roychoudhury, S., and Majumder, P. P. 2003. Ethnic India: A genomic view, with special reference to peopling and structure. *Genome Research*, 13, 2277–2290.

Batra, S. M. 1986. The sacredness of the cow in India. *Social Compass*, 33, 163–175.

Bayless, T. M., and Rosensweig, N. S. 1966. A racial difference in incidence of lactase deficiency: A survey of milk intolerance and lactase deficiency in healthy adult males. *Journal of the American Medical Association*, 197, 968–972.

————. 1967. Incidence and implications of lactase deficiency and milk intolerance in white and Negro populations. *Johns Hopkins Medical Journal*, 121, 54–64.

Bertron, P., Barnard, N. D., and Mills, M. 1999. Racial bias in federal nutrition policy, Part I: The public health implications of variations in lactase persistence. *Journal of the National Medical Association*, 91, 151–157.

Bhardwaj, S., Misra, A., Khurana, L., Gulati, S., Shah, P., and Vikram, N. K. 2008. Childhood obesity in Asian Indians: A burgeoning cause of insulin resistance, diabetes and sub-clinical inflammation. *Asia Pacific Journal of Clinical Nutrition*, 17 Suppl. 1, 172–175.

Bhishagratna, K. L. 1911. *The Sushruta Samhita: An English Translation Based on Original Sanskrit Text.* Vols. 1–3. Calcutta: Wilkins Press.

Bland, B. F. 2006. *Got Cow? Cattle in American Art 1820–2000.* Yonkers, NY: Hudson River Museum.

Bringeus, N.-A. 1992. A Swedish beer milk shake. In P. Lysaght (ed.), *Milk and Milk Products from Medieval to Modern Times.* Edinburgh: Canongate Press.

British Broadcasting Corporation. 2007. China drinks its milk. *BBC News*, 7 August.

Brown, G. R., Dickins, T. E., Sear, R. and Laland, K. N. 2011. Evolutionary accounts of human behavioural diversity. *Philosophical Transactions of the Royal Society B: Biological Sciences,* 366, 313–324.

Burger, J., Kirchner, M., Bramanti, B., Haak, W., and Thomas, M. G. 2007. Absence of the lactase-persistence-associated allele in early Neolithic Europeans. *Proceedings of the National Academy of Sciences,* 104, 3736–3741.

Burnett, J. 1999. *Liquid Pleasures: A Social History of Drinks in Modern Britain.* New York: Routledge.

Burton, D. 1993. *The Raj at Table.* Boston: Faber and Faber.

Calvert, K. 2003. Patterns of childrearing in America. In W. Koops and M. Zuckerman (eds.), *Beyond the Century of the Child: Cultural History and Developmental Psychology.* Philadelphia: University of Pennsylvania Press.

Carpenter, K. J. 2007. The work of Wallace Aykroyd: International nutritionist and author. *Journal of Nutrition,* 137, 873–878.

Case, A., and Paxson, C. 2008. Stature and status: Height, ability, and labor market outcomes. *Journal of Political Economy,* 116, 499–532.

Chandan, R. C. 2013. History and Consumption Trends. In R. C. Chandan and A. Kilara (eds.), *Manufacturing Yogurt and Fermented Milks.* 2nd ed. New York: John Wiley & Sons.

Chatterjee, P. 2001. *A Time for Tea: Women, Labor, and Post/Colonial Politics on an Indian Plantation.* Durham, NC: Duke University Press.

Chen, S., Lin, B.-Z., Baig, M., Mitra, B., Lopes, R. J., Santos, A. M., Magee, D. A., Azevedo, M., Tarroso, P., Sasazaki, S., Ostrowski, S., Mahgoub, O., Chaudhuri, T. K., Zhang, Y.-P., Costa, V., Royo, L. J., Goyache, F., Luikart, G., Boivin, N., Fuller, D. Q., Mannen, H., Bradley, D. G., and Beja-Pereira, A. 2010. Zebu cattle are an exclusive legacy of the South Asia Neolithic. *Molecular Biology and Evolution,* 27, 1–6.

Chiapparino, F. 1995. Milk and fondant chocolate and the emergence of the Swiss chocolate industry at the turn of the twentieth century. In M. R. Scharer and A. Fenton (eds.), *Food and Material Culture.* Edinburgh: Tuckwell Press.

Chutani, A. M. 2012. School lunch program in India: Background, objectives and components. *Asia Pacific Journal of Clinical Nutrition,* 21, 151–154.

Coe, S. D., Coe, M. D., and Huxtable, R. J. 1996. *The True History of Chocolate.* London: Thames and Hudson.

Collingham, Lizzie. 2006. *Curry: A Tale of Cooks and Conquerors.* New York: Oxford University Press.

Cooper, C., Westlake, S., Harvey, N., Javaid, K., Dennison, E., and Hanson, M. 2006. Review: Developmental origins of osteoporotic fracture. *Osteoporosis International,* 17, 337–347.

Courtiol, A., Raymond, M., Godelle, B., and Ferdy, J. B. 2010. Mate choice and human stature: Homogamy as a unified framework for understanding mating preferences. *Evolution,* 64, 2189–203.

Craig, L. A., Goodwin, B., and Grennes, T. 2004. The effect of mechanical refrigeration on nutrition in the United States. *Social Science History,* 28, 325–336.

Craig, O. E., Chapman, J., Heron, C., Willis, L. H., Bartosiewicz, L., Taylor, G., Whittle, A., and Collins, M. 2005. Did the first farmers of central and eastern Europe produce dairy foods? *Antiquity*, 79, 882–894.

Crooke, W. 1989 [1879]. *A Glossary of North Indian Peasant Life*. New York: Oxford University Press.

Crosby, A. W. 1986. *Ecological Imperialism: The Biological Expansion of Europe 900–1900*. New York: Cambridge University Press.

Crumbine, S. J., and Tobey, J. A. 1929. *The Most Nearly Perfect Food: The Story of Milk*. Baltimore, MD: Williams & Wilkins.

Cunningham, H. 1998. Histories of childhood. *American Historical Review*, 103, 1195–1208.

Deaton, A. 2007. Height, health, and development. *Proceedings of the National Academy of Sciences*, 104, 13232–13237.

———. 2008. Height, health, and inequality: The distribution of adult heights in India. *American Economics Review*, 98, 468–474.

Deaton, A., and Arora, R. 2009. Life at the top: The benefits of height. *Economics & Human Biology*, 7, 133–136.

De Beer, H. 2012. Dairy products and physical stature: A systematic review and meta-analysis of controlled trials. *Economics & Human Biology*, 10, 299–309.

Delgado, C. L. 2003. Rising consumption of meat and milk in developing countries has created a new food revolution. *Journal of Nutrition*, 133, 3907S–3910S.

Delgado, C. L., and Narrod, C. A. 2002. Impact of Changing Market Forces and Policies on Structural Change in the Livestock Industries of Selected Fast-Growing Developing Countries. Final Research Report of Phase I—Project on Livestock Industrialization, Trade and Social-Health-Environment Impacts in Developing Countries. Available: http://www.fao.org/wairdocs/LEAD/X6115E/x6115e00.HTM (accessed 23 May 2013).

Delgado, C. L., Narrod, C. A., and Tiongco, M. M. 2003. Impact of Changing Market Forces and Policies on Structural Change in the Livestock Industries of Selected Fast-Growing Developing Countries. Final Research Report of Phase II—Project on Livestock Industrialization, Trade and Social-Health-Environment Impacts in Developing Countries. Available: http://www.fao.org/WAIRDOCS/LEAD/X6170E/x6170e09.htm#bm09 (accessed 23 May 2013).

den Hartog, A. P. 1992. Modern nutritional problems and historical nutrition research, with special reference to the Netherlands. In H. J. Teuteberg (ed.), *European Food History: A Research Review*. Leicester: Leicester University Press.

———. 2001. Changing perceptions on milk as a drink in western Europe: The case of the Netherlands. In I. de Garine and V. de Garine (eds.), *Drinking: Anthropological Approaches*. New York: Berghahn Books.

Diener, P., Nonini, D., and Robkin, E. E. 1978. The dialectics of the sacred cow: Ecological adaptation versus political appropriation in the origins of India's cattle complex. *Dialectical Anthropology*, 3, 221–241.

Douglas, M. 2002 [1966]. *Purity and Danger: An Analysis of Concepts of Pollution and Taboo*. New York: Routledge.

Duffey, K. J., and Popkin, B. M. 2007. Shifts in patterns and consumption of beverages between 1965 and 2002. *Obesity*, 15, 2739–2747.

Dupras, T. L., Schwarcz, H. P., and Fairgrieve, S. I. 2001. Infant feeding and weaning practices in Roman Egypt. *American Journal of Physical Anthropology*, 115, 204–212.

DuPuis, E. M. 2002. *Nature's Perfect Food: How Milk Became America's Drink*. New York: New York University Press.

Durham, W. 1991. *Coevolution: Genes, Culture and Human Diversity*. Stanford, CA: Stanford University Press.

The Economist. 1936. India's food problem. *The Economist*, 125, 627–628.

Eden, T. 2006. *Cooking in America, 1590–1840*. Westport, CT: Greenwood Press.

Elliott, C. 2010. Eatertainment and the (re)classification of children's foods. *Food Culture and Society*, 13, 539–553.

———. 2011. "It's junk food and chicken nuggets": Children's perspectives on "kids' food" and the question of food classification. *Journal of Consumer Behaviour*, 10, 133–140.

Elwood, P. C., Pickering, J. E., Givens, D. I., and Gallacher, J. E. 2010. The consumption of milk and dairy foods and the incidence of vascular disease and diabetes: An overview of the evidence. *Lipids*, 45, 925–939.

Enattah, N. S., Sahi, T., Savilahti, E., Terwilliger, J. D., Peltonen, L., and Varvela, I. 2002. Identification of a variant associated with adult-type hypolactasia. *Nature Genetics*, 30, 233–237.

Eppright, E. S., and Swanson, P. P. 1955. Distribution of calories in diets of Iowa school children. *Journal of the American Dietetic Association*, 31, 144–148.

Fenton, A. 1992. Milk products in the everyday diet of Scotland. In P. Lysaght (ed.), *Milk and Milk Products from Medieval to Modern Times*. Edinburgh: Canongate Press.

Ferro-Luzzi, G. E. 1977. The logic of South Indian food offerings. *Anthropos*, 72, 529–556.

Fields, G. P. 2001. *Religious Therapeutics: Body and Health in Yoga, Ayurveda, and Tantra*. Albany: State University of New York Press.

Flatz, G. 1987. Genetics of lactose digestion in humans. *Advances in Human Genetics*, 16, 1–77.

Flatz, G., and Rotthauwe, H. W. 1973. Lactose nutrition and natural selection. *Lancet*, 2, 76–77.

Food and Agricultural Organization (FAO). 1981. World Food Program Terminal Evaluation Report on Project India 618 (Operation Flood I). Rome: FAO.

Freed, S. A., Freed, R. S., Ballard, R., Chattopadhyay, K., Diener, P., Dumont, L., Ferreira, J. V., Fuller, C. J., Harris, M., Lodrick, D. O., Malik, S. L., Mishra, S. N., Newell, W. H., Nonini, D. M., Odend'hal, S., Rajapurohit, A. R., Robkin, E. E., Sharma, U. M., Suryanarayana, M., and Verma, H. S. 1981. Sacred cows and water buffalo in India: The uses of ethnography [and comments and reply]. *Current Anthropology*, 22, 483–502.

Freidberg, S. 2009. *Fresh: A Perishable History.* Cambridge, MA: Harvard University Press.

Fuller, D. 2006. Agricultural origins and frontiers in South Asia: A working synthesis. *Journal of World Prehistory,* 20, 1–86.

Gallego Romero, I., Basu Mallick, C., Liebert, A., Crivellaro, F., Chaubey, G., Itan, Y., Metspalu, M., Eaaswarkhanth, M., Pitchappan, R., Villems, R., Reich, D., Singh, L., Thangaraj, K., Thomas, M. G., Swallow, D. M., Mirazón Lahr, M., and Kivisild, T. 2012. Herders of Indian and European cattle share their predominant allele for lactase persistence. *Molecular Biology and Evolution,* 29(1), 249–260.

Gandhi, M. K. 2001. *India of My Dreams.* Ahmedabad: Navajivan Publishing House.

Gandhi, M. K., and Kumarappa, B. 1954. *How to Serve the Cow.* Ahmedabad: Navajivan Publishing House.

George, S. 1985. *Operation Flood: An Appraisal of Current Indian Dairy Policy.* Delhi: Oxford University Press.

Gerbault, P., Moret, C., Currat, M., and Sanchez-Mazas, A. 2009. Impact of selection and demography on the diffusion of lactase persistence. *PLoS One,* 4, e6369.

Ghatwai, M. 2008. Buffalo milk vs cow milk, MP has a fat problem. *Indian Express,* 11 August.

Godfray, H. C. J., Beddington, J. R., Crute, I. R., Haddad, L., Lawrence, D., Muir, J. F., Pretty, J., Robinson, S., Thomas, S. M., and Toulmin, C. 2010. Food security: The challenge of feeding 9 billion people. *Science,* 327, 812–818.

Goldschmidt, B. 2012. Value-added dairy: Milk money. *Progressive Grocer.* Available: http://www.progressivegrocer.com/top-stories/headlines/cpgs-trading-partners/id36921/value-added-dairy-milk-money/ (accessed 30 November 2013).

Gopalan, C., Rama Sastri, B. V., and Balasubramanian, S. C. 1984. *Nutritive Value of Indian Foods.* Hyderabad, India: National Institute of Nutrition, Indian Council of Medical Research.

Gopalan, H., and Ramachandran, P. 2011. Assessment of nutritional status in Indian preschool children using WHO 2006 growth standards. *Indian Journal of Medical Research,* 134, 47–53.

Gould, W. 2004. *Hindu Nationalism and the Language of Politics in Late Colonial India.* New York: Cambridge University Press.

Grigg, D. 1999. The fat of the land: A geography of oil and fat consumption. *GeoJournal,* 48, 259–268.

———. 2002. The worlds of tea and coffee: Patterns of consumption. *GeoJournal,* 57, 283–294.

Grillenberger, M., Neumann, C. G., Murphy, S. P., Bwibo, N. O., Van't Veer, P., Hautvast, J. G. A. J., and West, C. E. 2003. Food supplements have a positive impact on weight gain and the addition of animal source foods increases lean body mass of Kenyan schoolchildren. *Journal of Nutrition,* 133, 3957S–3964S.

Guha, A. 2006. Ayurvedic concept of food and nutrition. *SoM Articles,* Paper 25. Available: http://digitalcommons.uconn.edu/som_articles/25 (accessed 30 November 2013).

Gunderson, G. W. 1971. National School Lunch Program (NSLP): Background and Development. U.S. Department of Agriculture, Food and Nutrition Service. Available: http://www.fns.usda.gov/nslp/history (accessed 30 November 2013).

Gupta, C. 2001. The icon of mother in late colonial North India: "Bharat Mata," "Matri Bhasha" and "Gau Mata." *Economic and Political Weekly,* 36, 4291–4299.

Halpern, S. A. 1988. *American Pediatrics: The Social Dynamics of Professionalism, 1880–1980.* Berkeley: University of California Press.

Hamilton, F. 1807. A journey from Madras through the countries of Mysore, Canara, and Malabar, performed under the orders of the Most Noble the Marquis Wellesley, Governor General of India: For the express purpose of investigating the state of agriculture, arts, and commerce; the religion, manners, and customs; the history, natural and civil, and antiquities, in the dominions of the rajah of Mysore, and the countries acquired by the Honourable East India Company, in the late and former wars, from Tippoo Sultaun. London: T. Cadell and W. Davies.

Harner, M. 1977. The enigma of Aztec sacrifice. *Natural History,* 86, 46–51.

Harris, M. 1966. The cultural ecology of India's sacred cattle. *Current Anthropology,* 7, 51–66.

———. 1979. *Cultural Materialism: The Stuggle for a Science of Culture.* Walnut Creek, CA: AltaMira Press.

———. 1985. *Good to Eat: Riddles of Food and Culture.* Prospect Heights, IL: Waveland Press.

———. 1989. *Cows, Pigs, Wars, and Witches: The Riddles of Culture.* New York: Vintage.

Hartley, R. M. 1977 [1842]. *An Historical, Scientific and Practical Essay on Milk as an Article of Human Sustenance.* New York: Arno Press.

Heaney, R. P. 2001. The dairy controversy: Facts, questions, and polemics. In P. Burckhardt, B. Dawson-Hughes, and R. P. Heaney (eds.), *Nutritional Aspects of Osteoporosis.* New York: Academic Press.

Heaney, R. P., and Weaver, C. M. 1990. Calcium absorption from kale. *American Journal of Clinical Nutrition,* 51, 656–657.

Hoag, C. 2011. Schools may ban chocolate milk over added sugar. *USA Today,* 11 May.

Hoffpauir, R. 1977. The Indian milk buffalo: A paradox of high performance and low reputation. *Asian Profile,* 5, 111–134.

———. 1982. The water buffalo: India's other bovine. *Anthropos,* 77, 215–238.

Hoppe, C., Mølgaard, C., and Michaelsen, K. F. 2006. Cow's milk and linear growth in industrialized and developing countries. *Annual Review of Nutrition,* 26, 131–173.

Huang, H. T. 2002. Hypolactasia and the Chinese diet. *Current Anthropology,* 43, 809–819.

Huard, R. C., Fabre, M., De Haas, P., Claudio Oliveira Lazzarini, L., Van Soolingen, D., Cousins, D., and Ho, J. L. 2006. Novel genetic polymorphisms that further delineate the phylogeny of the mycobacterium tuberculosis complex. *Journal of Bacteriology,* 188, 4271–4287.

India Study Team. 2007. The dairy industry of India. *Dairy Updates: World Dairy Industries,* no. 10. Available: http://babcock.wisc.edu/sites/default/files/documents/productdownload/du_108.en_.pdf (accessed 28 April 2013).

Ingram, C. J., Elamin, M. F., Mulcare, C. A., Weale, M. E., Tarekegn, A., Raga, T. O., Bekele, E., Elamin, F. M., Thomas, M. G., Bradman, N., Swallow, D. M., Itan, Y., Tishkoff, S. A., Reed, F. A., Ranciaro, A., Voight, B. F., Babbitt, C. C., Silverman, J. S., Powell, K., Mortensen, H. M., Hirbo, J. B., Osman, M., Ibrahim, M., Omar, S. A., Lema, G., Nyambo, T. B., Ghori, J., Bumpstead, S., Pritchard, J. K., Wray, G. A., and Deloukas, P. 2007. A novel polymorphism associated with lactose tolerance in Africa: Multiple causes for lactase persistence? *Human Genetics,* 120, 779–788.

Ingram, C. J., Mulcare, C. A., Itan, Y., Thomas, M. G., and Swallow, D. M. 2009. Lactose digestion and the evolutionary genetics of lactase persistence. *Human Genetics,* 124, 579–591.

Itan, Y., Powell, A., Beaumont, M. A., Burger, J., and Thomas, M. G. 2009. The origins of lactase persistence in Europe. *PLoS Computational Biology,* 5, e1000491.

IUF Dairy Division. 2011. Indian dairy industry. Available: http://cms.iuf.org/sites/cms.iuf.org/files/Indian%20Dairy%20Industry.pdf (accessed 23 May 2013).

Jarvis, J. K., and Miller, G. D. 2002. Overcoming the barrier of lactose intolerance to reduce health disparities. *Journal of the National Medical Association,* 94, 55–66.

Jha, D. N. 2002. *The Myth of the Holy Cow.* London: Verso.

Jing, J. 2000. Introduction: Food, children, and social change in contemporary China. In J. Jing (ed.), *Feeding China's Little Emperors: Food, Children, and Social Change.* Stanford, CA: Stanford University Press.

Kanjilal, B., Mazumdar, P., Mukherjee, M., and Rahman, M. H. 2010. Nutritional status of children in India: Household socio-economic condition as the contextual determinant. *International Journal for Equity in Health,* 9, 19.

Khurody, D. N. 1974. *Dairying in India: A Review.* New York: Asia Publishing House.

Komlos, J., and Baur, M. 2004. From the tallest to (one of) the fattest: The enigmatic fate of the American population in the 20th century. *Economics & Human Biology,* 2, 57–74.

Komlos, J., and Lauderdale, B. E. 2007. Spatial correlates of US heights and body mass indexes, 2002. *Journal of Biosocial Science,* 39, 59–78.

Korom, F. J. 2000. Holy cow! The apotheosis of zebu, or why the cow is sacred in Hinduism. *Asian Folklore Studies,* 59, 181–203.

Kurien, V. 1997. *An Unfinished Dream.* New York: Tata McGraw-Hill.

————. 2007. India's milk revolution: Investing in rural producer associations. In D. Narayan and E. Glinskaya (eds.), *Ending Poverty in South Asia: Ideas That Work.* Washington, DC: World Bank.

Kutumbiah, P. 1959. Pediatrics (Kaumara bhrtya) in ancient India. *Indian Journal of Pediatrics,* 26, 328–337.

Lacomb, R. P., Sebastian, R. S., Wilkinson Enns, C., and Goldman, J. D. 2011. Beverage choices of U.S. adults: What we eat in America, NHANES 2007–2008. *Food Surveys Research Group Dietary Data Brief No. 6.* Available: http://www.ars.usda.gov/SP2UserFiles/Place/12355000/pdf/DBrief/6_beverage_choices_adults_0708.pdf (accessed 25 November 2013).

Lanou, A. J. 2009. Should dairy be recommended as part of a healthy vegetarian diet? Counterpoint. *American Journal of Clinical Nutrition,* 89, 1638S–1642S.

Lanou, A. J., Berkow, S. E., and Barnard, N. D. 2005. Calcium, dairy products, and bone health in children and young adults: A reevaluation of the evidence. *Pediatrics,* 115, 736–743.

Lee, G. C. 1900. *Leading Documents of English History.* New York: Henry Holt and Company.

Leighton, G., and Clark, M. L. 1929. Milk consumption and the growth of schoolchildren. *Lancet,* i, 40–43.

Levenstein, H. A. 1988. *Revolution at the Table: The Transformation of the American Diet.* New York: Oxford University Press.

————. 2012. *Fear of Food: A History of Why We Worry about What We Eat.* Chicago: University of Chicago Press.

Lewitt, E. M., and Kerrebrock, N. 1997. Population-based growth stunting. *The Future of Children,* 7, 149–156.

Lodrick, D. O. 1981. *Sacred Cows, Sacred Places: Origins and and Survival of Animal Homes in India.* Berkeley: University of California Press.

Lutgendorf, P. 2012. Making tea in India: Chai, capitalism, culture. *Thesis Eleven,* 113, 11–31.

Lysaght, P. (ed.). 1992. *Milk and Milk Products from Medieval to Modern Times.* Edinburgh: Canongate Press.

MacFarlane, A., and MacFarlane, I. 2004. *The Empire of Tea: The Remarkable History of the Plant That Took Over the World.* Woodstock, NY: Overlook Press.

Macintosh, W. 1782. Travels in Europe, Asia, and Africa; describing characters, customs, manners, laws, and productions of nature and art: Containing various remarks on the political and commercial interests of Great Britain: And delineating, in particular, a new system for the government and improvement of the British settlements in the East Indies: Begun in the year 1777, and finished in 1781. London: J. Murray.

MacLachlan, M. D. 1982. Bovine sex and species ratios in India [comments]. *Current Anthropology,* 23, 376–377.

Mahias, M.-C. 1988. Milk and its transmutations in Indian Society. *Food and Foodways,* 2, 265–288.

Mair, V. H., and Hoh, E. 2009. *The True History of Tea*. New York: W. W. Norton.

Malamoud, C. 1998. *Cooking the World: Ritual and Thought in Ancient India*. New York: Oxford University Press.

Malik, V. S., Popkin, B. M., Bray, G. A., Després, J.-P., and Hu, F. B. 2010. Sugar-sweetened beverages, obesity, type 2 diabetes mellitus, and cardiovascular disease risk. *Circulation*, 121, 1356–1364.

Markowitz, D. L., and Cosminsky, S. 2005. Overweight and stunting in migrant Hispanic children in the USA. *Economics & Human Biology*, 3, 215–240.

Marriott, B. M., Campbell, L., Hirsch, E., and Wilson, D. 2007. Preliminary data from demographic and health surveys on infant feeding in 20 developing countries. *Journal of Nutrition*, 137, 518S–523S.

McCollum, E. V. 1922. *The Newer Knowledge of Nutrition*. New York: Macmillan.

———. 1957. *A History of Nutrition*. Boston: Houghton Mifflin.

McDowell, M. A., Fryar, C. D., Ogden, C. L., and Flegal, K. M. 2008. Anthropometric reference data for children and adults: United States, 2003–2006. *National Health Statistics Reports*, No. 10. Available: www.cdc.gov/nchs/data/nhsr/nhsr010.pdf (accessed 23 May 2013).

Meadow, R. H. 1996. The origins and spread of agriculture and pastoralism in northwest South Asia. In D. R. Harris (ed.), *The Origins and Spread of Agriculture and Pastoralism in Eurasia*. London: Psychology Press.

Mendelson, A. 2008. *Milk: The Surprising Story of Milk through the Ages*. New York: Alfred A. Knopf.

Midha, T., Nath, B., Kumari, R., Rao, Y., and Pandey, U. 2012. Childhood obesity in India: A meta-analysis. *Indian Journal of Pediatrics*, 79, 945–948.

Milk Processors Education Program. 2009. The New Face of Wellness. MilkPEP. Available: http://www.whymilk.com/new_face_of_wellness.php (accessed 1 June 2009).

———. n.d. *Got Milk? Get Tall!* [Online]. Available: http://www.whymilk.com/facts_gotmilk.htm [accessed 20 September 2006].

Miller, B. D. 1987. Female infanticide and child neglect in rural north India. In N. Scheper-Hughes (ed.), *Child Survival: Anthropological Perspectives on the Treatment and Maltreatment of Children*. Boston: D. Reidel Publishing Company.

Mintz, S. 1985. *Sweetness and Power: The Place of Sugar in Modern History*. New York: Viking.

———. 1996. *Tasting Food, Tasting Freedom: Excursions into Eating, Culture, and the Past*. Boston: Beacon Press.

Mishra, Y., Panda, G. R., Gonsalves, C. 2009. *Human Rights and Budgets in India*. New Delhi: Human Rights Law Network.

Mitra, N. 2007. A drop of milk. In J. Thieme and I. Raja (eds.), *The Table Is Laid: The Oxford Anthology of South Asian Food Writing*. New Delhi: Oxford University Press.

Moffet, T., James, R., and Oldys, W. 2011 [1746]. *Health's Improvement Or, Rules Comprizing and Discovering the Nature, Method and Manner of Pre-*

paring All Sorts of Foods Used in This Nation. London: T. Osborne in Gray's Inn.

Moxham, R. 2003. *Tea: Adaptation, Exploitation and Empire.* New York: Carroll & Graf Publishers.

Mullaly, J. 1853. *The Milk Trade of New York and Vicinity, Giving an Account of the Sale of Pure and Adulterated Milk.* New York: Fowlers and Wells.

Murasko, J. E. 2013. Physical growth and cognitive skills in early-life: Evidence from a nationally representative US birth cohort. *Social Science & Medicine,* 97, 267–277.

Nanda, A. S., and Nakao, T. 2003. Role of buffalo in the socioeconomic development of rural Asia: Current status and future prospectus. *Animal Science Journal,* 74, 443–455.

National Dairy Council. 2012. Lactose intolerance among different ethnic groups. Available: http://www.nationaldairycouncil.org/SiteCollectionDocuments /LI%20and%20Minorites_FINALIZED.pdf (accessed 23 May 2013).

National Dairy Development Board. 2012. Per capita monthly consumption expenditure in milk & milk products. Available: http://www.nddb.org/English /Statistics/Pages/Expenditure-Milk.aspx (accessed 29 February 2012).

National Institute of Child Health & Human Development. n.d. Milk matters. Available: http://www.nichd.nih.gov/milk (accessed 23 May 2013).

National Institute of Nutrition. 2010. Dietary guidelines for Indians. Available: http://www.indg.gov.in/health/nutrition/dietary-guidelines-for-indians (accessed 23 May 2013).

National Nutrition Monitoring Bureau. 2002. Diet and nutritional status of rural population. NNMB Technical Report No. 21. Hyderabad, India: National Institute of Nutrition.

Nestle, M. 1998. Toward more healthful dietary patterns: A matter of policy. *Public Health Reports,* 113, 420–423.

———. 2002. *Food Politics: How the Food Industry Influences Nutrition and Health.* Berkeley: University of California Press.

Nettle, D. 2002. Women's height, reproductive success and the evolution of sexual dimorphism in modern humans. *Proceedings of the Royal Society of London. Series B: Biological Sciences,* 269, 1919–1923.

Nicklas, T. A. 2003. Calcium intake trends and health consequences from childhood through adulthood. *Journal of the American College of Nutrition,* 22, 340–356.

NIH Consensus Development Conference. 2010. Lactose intolerance and health: Final panel statement. Available: http://consensus.nih.gov/2010/lactosestate ment.htm (accessed 23 May 2013).

O'Flaherty Doniger, W. 1980. *Women, Androgynes, and Other Mythical Beasts.* Chicago: University of Chicago Press.

Oftedal, O. T., and Iverson, S. J. 1995. Comparative analysis of nonhuman milks. In R. G. Jensen (ed.), *Handbook of Milk Composition.* New York: Academic Press.

Olivelle, P. 2002. Food for thought: Dietary regulations and social organization in ancient India. 2001 Gonda Lecture. Amsterdam: Royal Netherlands Academy of Arts and Sciences.

Orr, J. B. 1928. Milk consumption and the growth of school-children. *Lancet*, i, 202–203.

Ortiz de Montellano, B. R. 1978. Aztec cannibalism: An ecological necessity? *Science*, 200, 611–617.

Paige, D. M., Bayless, T. M., and Graham, G. G. 1972. Milk programs: Helpful or harmful to Negro children? *American Journal of Public Health*, 62, 1486–1488.

Parameswaran, R. 2011. States of imagination: Aesthetics, affects and representational practices in/of Asia. Paper presented at the American Folklore Society Conference, Indiana University, Bloomington, IN.

Patton, S. 2004. *Milk: Its Remarkable Contribution to Human Health and Well-Being*. New Brunswick, NJ: Transaction Publishers.

Paul, E. J. 2001. *The Story of Tea*. New Delhi: Roli Books.

Pawlowski, B., Dunbar, R. I. M., and Lipowicz, A. 2000. Evolutionary fitness: Tall men have more reproductive success. *Nature*, 403, 156–156.

Pendergrast, M. 1999. *Uncommon Grounds: The History of Coffee and How It Transformed Our World*. New York: Basic Books.

Pollock, J. 2006. Two controlled trials of supplementary feeding of British school children in the 1920s. *Journal of the Royal Society of Medicine*, 99, 323–327.

Popkin, B. M. 2010. Patterns of beverage use across the lifecycle. *Physiology & Behavior*, 100, 4–9.

Popkin, B. M., Adair, L. S., and Ng, S. W. 2012. Global nutrition transition and the pandemic of obesity in developing countries. *Nutrition Reviews* 70, 3–21.

Popkin, B. M., and Nielsen, S. J. 2003. The sweetening of the world's diet. *Obesity Research*, 11, 1325–1332.

Prakash, O. 1987. *Economy and Food in Ancient India*. Delhi: Bharatiya Vidya Prakashan.

Prowse, T., Saunders, S., Fitzgerald, C., Bondioli, L., and Macchiarelli, R. 2010. Growth, morbidity, and mortality in antiquity: A case study from Imperial Rome. In T. Moffat and T. Prowse (eds.), *Human Diet and Nutrition in Biocultural Perspective: Past Meets Present*, pp. 173–196. New York: Berghahn Books.

Rai, A. R. 2006. Mother Dairy's BIG India plans. *Rediff News*, 4 July.

Ramachandran, P. 2007. Nutrition transition in India 1947–2007. Available: http://wcd.nic.in/research/nti1947/7.2%20dietary%20intakes%20pr%20 4.2.pdf (accessed 23 May 2013).

Rand, H. 1998. Wat maakte de "Keukenmeid" van Vermeer? *Bulletin van het Rijksmuseum*, 46, 275–278.

Reaney, B. C. 1922. Milk and our School Children. Prepared for the Bureau of Education by the Child Health Organization of America. Washington, DC: Government Printing Office.

Richey, H. G. 1937. The relation of accelerated, normal and retarded puberty to the height and weight of school children. *Monographs of the Society for Research in Child Development*, 2, i–67.

Rifkind, H. R. 2007. Fresh foods for the Army, 1775–1950. Available: http://www
.qmfound.com/fresh_foods_for_the_army_1775_1950.htm (accessed 23
May 2013).

Riley, G. 2004. Images of infant nutrition: Sightings of food in group & child
portraits. In R. Hosking (ed.), *Nurture: Proceedings of the Oxford Sympo-
sium on Food and Cookery 2003.* Bristol, UK: Footwork.

Rorabaugh, W. J. 1976. Estimated U.S. alcoholoic beverage consumption, 1790–
1860. *Journal of Studies on Alcohol and Drugs,* 37, 357–364.

Rosensweig, N. S. 1973. Lactose feeding and lactase deficiency. *American Journal
of Clinical Nutrition,* 26, 1166–1167.

Ross, A. C., Manson, J. E., Abrams, S. A., Aloia, J. F., Brannon, P. M., Clinton,
S. K., Durazo-Arvizu, R. A., Gallagher, J. C., Gallo, R. L., Jones, G., Kovacs,
C. S., Mayne, S. T., Rosen, C. J., and Shapses, S. A. 2011. The 2011 report on
dietary reference intakes for calcium and vitamin D from the Institute of
Medicine: What clinicians need to know. *Journal of Clinical Endocrinology
& Metabolism,* 96, 53–58.

Roth, S., Gloy, A., Hyde, J., and Kelly, B. 2008. Get more from your milk: Increasing profits through value-added products. Pennsylvania State University. Available: http://extension.psu.edu/pubs/xa0019 (accessed 23 May
2013).

Rusoff, L. 1955. The miracle of milk. *Journal of Dairy Science,* 38, 1057.

Saberi, H. 2010. *Tea: A Global History.* London: Reaktion Books.

Sahi, T. 1994a. Genetics and epidemiology of adult-type hypolactasia. *Scandinavian Journal of Gastroenterology,* 29, 7–20.

———. 1994b. Hypolactasia and lactase persistence: Historical review and the
terminology. *Scandinavian Journal of Gastroenterology,* 29, 1–6.

Sahlins, M. 1976. *Culture and Practical Reason.* Chicago: University of Chicago
Press.

Saxena, R. 1996. Demand for milk and milk products in India. *Institute of Rural
Management, Anand,* Working Paper 103.

Scrinis, G. 2013. *Nutritionism: The Science and Politics of Dietary Advice.* New
York: Columbia University Press.

Sear, R., and Marlowe, F. W. 2009. How universal are human mate choices? Size
does not matter when Hadza foragers are choosing a mate. *Biology Letters,*
5, 606–609.

Sebastian, R., Goldman, J., Wilkinson Enns, C., and Lacomb, R. 2010. Fluid milk
consumption in the United States: What we eat in America, NHANES 2005–
2006. *Food Surveys Research Group Dietary Data Brief,* No. 3. Available:
http://www.ars.usda.gov/SP2UserFiles/Place/12355000/pdf/DBrief/3_milk
_consumption_0506.pdf (accessed 25 November 2013).

Sen, C. T. 2004. *Food Culture in India.* Westport, CT: Greenwood Press.

Seymour, S. C. 1999. *Women, Family, and Child Care in India: A World in Transition.* New York: Cambridge University Press.

Shammas, C. 1990. *The Pre-industrial Consumer in England and America.* Oxford: Clarendon Press.

Sharma, R., Grover, V., and Chaturvedi, S. 2011. Recipe for diabetes disaster: A study of dietary behaviors among adolescent students in South Delhi, India. *International Journal of Diabetes in Developing Countries,* 31, 4–8.

Sharma, V. P., Singh, R. V., Staal, S., and Delgado, C. L. 2002. *Annex I: Critical Issues for Poor People in the Indian Dairy Sector on the Threshold of a New Era.* Rome: Food and Agriculture Organization.

Shetty, P. 2012. Public health: India's diabetes time bomb. *Nature,* 485, S14–S16.

Shetty, P. S. 2002. Nutrition transition in India. *Public Health Nutrition,* 5, 175–182.

Simoons, F. J. 1970. The traditional limits of milking and milk use in Southern Asia. *Anthropos,* 65, 547–593.

———. 1978. The geographic hypothesis and lactose malabsorption. *American Journal of Digestive Diseases,* 23, 963–980.

———. 1979. Questions in the sacred-cow controversy. *Current Anthropology,* 20, 467–493.

———. 1981. Geographic patterns of primary adult lactose malabsorption: A further interpretation of evidence for the Old World. In D. M. Paige and T. M. Bayless (eds.), *Lactose Digestion: Clinical and Nutritional Implications.* Baltimore: Johns Hopkins University Press.

———. 2001. Persistence of lactase activity among northern Europeans: A weighing of evidence for the calcium absorption hypothesis. *Ecology of Food and Nutrition,* 40, 397–469.

Singh, G. 2012. Warning: Bulk of milk supplied by vendors across the country is either contaminated or adulterated, reveals govt report. *India Today,* 20 October.

Singh, R. 2011. India Dairy and Products Annual, USDA Foreign Agricultural Service Gain Report. Available: http://agriexchange.apeda.gov.in/marketreport/Reports/India_Dairy_report.pdf (accessed 23 May 2013).

Sinha, R. P. 1961. *Food in India: An Analysis of the Prospects for Self-Sufficiency by 1975–1976.* New York: Oxford University Press.

Smith, A. F. (ed.). 2004. *The Oxford Encyclopedia of Food and Drink in America.* New York: Oxford University Press.

Smits, J., and Monden, C. W. S. 2012. Taller Indian women are more successful at the marriage market. *American Journal of Human Biology,* 24, 473–478.

Spears, D. 2012. Height and cognitive achievement among Indian children. *Economics & Human Biology,* 10, 210–219.

Srinivas, M. N. 1956. A note on Sanskritization and Westernization. *Journal of Asian Studies,* 15, 481–496.

Stavely, K., and Fitzgerald, K. 2004. *America's Founding Food: The Story of New England Cooking.* Chapel Hill: University of North Carolina Press.

Steckel, R. H. 2009. Heights and human welfare: Recent developments and new directions. *Explorations in Economic History,* 46, 1–23.

Steel, F. A., and Gardiner, G. 2010 [1898]. *The Complete Indian Housekeeper and Cook.* New York: Oxford University Press.

Stulp, G., Verhulst, S., Pollet, T. V., and Buunk, A. P. 2012. The effect of female height on reproductive success is negative in Western populations, but more variable in non-Western populations. *American Journal of Human Biology,* 24, 486–494.

Subramanian, S. V., Özaltin, E., and Finlay, J. E. 2011. Height of nations: A socio-economic analysis of cohort differences and patterns among women in 54 low- to middle-income countries. *PLoS One,* 6, e18962.

Tanner, J. M. 1982. The potential of auxological data for monitoring economic and social well-being. *Social Science History,* 6, 571–581.

Taubes, G. 2001. The soft science of dietary fat. *Science,* 291, 2536–2545.

Thapar, R. 2003. *Early India: From the Origins to AD 1300.* Berkeley: University of California Press.

Tishkoff, S. A., Reed, F. A., Ranciaro, A., Voight, B. F., Babbitt, C. C., Silverman, J. S., Powell, K., Mortensen, H. M., Hirbo, J. B., Osman, M., Ibrahim, M., Omar, S. A., Lema, G., Nyambo, T. B., Ghori, J., Bumpstead, S., Pritchard, J. K., Wray, G. A., and Deloukas, P. 2007. Convergent adaptation of human lactase persistence in Africa and Europe. *Nature Genetics,* 39, 31–40.

Uberoi, P. 2003. "Unity in diversity?" Dilemmas of nationhood in Indian calendar art. In S. Ramaswamy (ed.), *Beyond Appearances? Visual Practices and Ideologies in Modern India,* pp. 191–232. Thousand Oaks, CA: Sage Publications.

U.S. Department of Agriculture, Economic Research Service. 2013. ERS dairy consumption trends. Available: http://www.ers.usda.gov/data-products/dairy-data .aspx (accessed 23 May 2013).

U.S. Department of Commerce. 2002. *Statistical Abstract of the United States: 2002.* Washington, DC: United States Census.

U.S. Department of Health and Human Services and U.S. Department of Agriculture. 2010. Dietary guidelines for Americans. Available: http://health.gov /dietaryguidelines/dga2010/DietaryGuidelines2010.pdf (accessed 17 August 2011).

Vaidyanathan, A., Nair, K. N., Harris, M., Abruzzi, W. S., Adams, R. N., Batra, S. M., Chibnik, M., Crotty, R., Doherty, V. S., Freed, S. A., Freed, R. S., Fruehling, R. T., Maclachlan, M. D., Nonini, D. M., Odend'hal, S., Rao, D. L. P., and Robkin, E. 1982. Bovine sex and species ratios in India [and comments and reply]. *Current Anthropology,* 23, 365–383.

Valenze, D. 2011. *Milk: A Local and Global History.* New Haven, CT: Yale University Press.

Van Winter, J. M. 1992. The consumption of dairy products in the Netherlands in the fifteenth and sixteenth centuries. In P. Lysaght (ed.), *Milk and Milk Products from Medieval to Modern Times.* Edinburgh: Canongate Press.

Velten, H. 2010. *Milk: A Global History.* London: Reaktion Books.

Vijayagopalan, S. 1988. *Domestic Marketing of Tea: A Reappraisal.* New Delhi: National Council of Applied Economic Research.

Vijayapushpam, T., Menon, K. K., Rao, D. R., and Antony, G. M. 2003. A qualitative assessment of nutrition knowledge levels and dietary intake of schoolchildren in Hyderabad. *Public Health Nutrition,* 6, 683–688.

Walvin, J. 1997. *Fruits of Empire: Exotic Produce and British Taste, 1660–1800.* London: Macmillan.

Wattiaux, M. A. 2011. Milk composition and nutritional value. In *Dairy Essentials*. Madison, WI: Babcock Institute for International Dairy Research and Development.

Weaver, C. M., Heaney, R. P., Nickel, K. P., and Packard, P. I. 1997. Calcium bioavailability from high oxalate vegetables: Chinese vegetables, sweet potatoes and rhubarb. *Journal of Food Science, 62*, 524–525.

Welsh, S. O., Davis, C., and Shaw, A. 1993. *USDA's Food Guide: Background and Development*. U.S. Department of Agriculture, Human Nutrition Information Service, Miscellaneous Publication No. 1514.

White, A. F. 2009. Performing the promise of plenty in the USDA's 1933–34 World's Fair exhibits. *Text and Performance Quarterly, 29*, 22–43.

Wiley, A. S. 1992. Adaptation and the biocultural paradigm in medical anthropology: A critical review. *Medical Anthropology Quarterly, 6*, 216–236.

———. 1993. Evolution, adaptation, and the role of biocultural medical anthropology. *Medical Anthropology Quarterly, 7*, 192–199.

———. 2004. "Drink milk for fitness": The cultural politics of human biological variation and milk consumption in the United States. *American Anthropologist, 106*, 506–517.

———. 2005. Does milk make children grow? Relationships between milk consumption and height in NHANES 1999–2002. *American Journal of Human Biology, 17*, 425–441.

———. 2007. Transforming milk in a global economy. *American Anthropologist, 109*, 666–677.

———. 2011a. Milk for "growth": Global and local meanings of milk consumption in China, India, and the U.S. *Food and Foodways, 19*, 11–33.

———. 2011b. Milk intake and total dairy consumption: Associations with early menarche in NHANES 1999–2004. *PLoS One, 6*, e14685.

———. 2011c. *Re-imagining Milk*. New York: Routledge.

———. 2012. Cow milk consumption, insulin-like growth factor-I, and human biology: A life history approach. *American Journal of Human Biology, 24*, 130–138.

Wiley, A. S., and Allen, J. S. 2013. *Medical Anthropology: A Biocultural Approach*. 2nd ed. New York: Oxford University Press.

Wilson, C. A. 1991. *Food and Drink in Britain: From the Stone Age to the 19th Century*. Chicago: Academy Chicago Publishers.

Working Group on Children under Six. 2007. Strategies for children under six. *Economic and Political Weekly, 42*, 87–101.

Wright, N. C. 1937. *Report on the Development of the Cattle and Dairy Industries of India*. Delhi: Government of India Press.

Wujastyk, D. 2001. *The Roots of Ayurveda: Selections from the Ayurvedic Classics*. New York: Penguin.

Yang, S., Tilling, K., Martin, R., Davies, N., Ben-Shlomo, Y., and Kramer, M. S. 2011. Pre-natal and post-natal growth trajectories and childhood cognitive ability and mental health. *International Journal of Epidemiology, 40*, 1215–1226.

Yang, X. G., Li, Y. P., Ma, G. S., Hu, X. Q., Wang, J. Z., Cui, Z. H., Wang, Z. H., Yu, W. T., Yang, Z. X., and Zhai, F. Y. 2005. [Study on weight and height of the Chinese people and the differences between 1992 and 2002]. *Zhonghua Liu Xing Bing Xue Za Zhi*, 26, 489–493.

Yeoman, B. 2003. Is the US government making children fat? *Nieman Reports*, 57, 30.

Ziegler, E. E. 2007. Adverse effects of cow's milk in infants. *Nestle Nutrition Workshop Series: Pediatric Program*, 60, 185–199.

Index